Green Infrastructure

Our understandings of the landscapes around us are constantly changing. How we interact with, manage and value these spaces is important, as it helps us to ensure we live in attractive, functional and sustainable places. Green Infrastructure planning is the current 'go-to' approach in landscape planning that incorporates human–environment interactions, understandings of ecology and how sociocultural factors influence our use of parks, gardens and waterways.

This book explores several interpretations of Green Infrastructure bringing together case studies of policy, practice, ecological change and community understandings of landscape. Focusing on how planning policy shapes our interactions with the landscape, as individuals and communities, the book discusses what works and what needs to be improved. It examines how environmental management can promote more sustainable approaches to landscape protection ensuring that water resources and ecological communities are not harmed by development. It also asks what the economic and community values of Green Infrastructure are to illustrate how different social, ecological and political factors influence how our landscapes are managed.

The central message of the book focusses on the promotion of multifunctional nature within urban landscapes that helps people, the economy and the environment to meet the challenges of population, infrastructure and economic change.

The chapters in this book were originally published as a special issue in *Landscape Research*.

Ian C. Mell is a Lecturer in Environmental and Landscape Planning at University of Manchester, UK. He researches human–environment interactions exploring social, economic and ecological valuation of Green Infrastructure in the UK, EU, USA and Asia. He is the author of *Global Green Infrastructure: Lessons for Successful Policy-Making, Investment and Management* (2016, Routledge).

Green Infrastructure

Current Debates for Policy, Practice and Implementation

Edited by
Ian C. Mell

Routledge
Taylor & Francis Group

LONDON AND NEW YORK

First published 2018 by Routledge

2 Park Square, Milton Park, Abingdon, Oxfordshire OX14 4RN
52 Vanderbilt Avenue, New York, NY 10017

Routledge is an imprint of the Taylor & Francis Group, an informa business

First issued in paperback 2019

British Library Cataloguing in Publication Data
A catalogue record for this book is available from the British Library

ISBN 13: 978-1-138-56119-9 (hbk)
ISBN 13: 978-0-367-89222-7 (pbk)

Typeset in Myriad Pro
by diacriTech, Chennai

Publisher's Note
The publisher accepts responsibility for any inconsistencies that may have arisen
during the conversion of this book from journal articles to book chapters, namely
the possible inclusion of journal terminology.

Disclaimer
Every effort has been made to contact copyright holders for their permission to
reprint material in this book. The publishers would be grateful to hear from any
copyright holder who is not here acknowledged and will undertake to rectify any
errors or omissions in future editions of this book.

Contents

CONTENTS

Citation Information

The following chapters in this book were originally published in *Landscape Research*, volume 42, issue 2 (February 2017). When citing this material, please use the original page numbering for each article, as follows:

Chapter 1
Green infrastructure: reflections on past, present and future praxis
Ian C. Mell
Landscape Research, volume 42, issue 2 (February 2017) pp. 135–145

Chapter 2
The emergence of green infrastructure as promoting the centralisation of a landscape perspective in spatial planning—the case of Ireland
Mick Lennon, Mark Scott, Marcus Collier and Karen Foley
Landscape Research, volume 42, issue 2 (February 2017) pp. 146–163

Chapter 3
Urban green infrastructure and urban forests: a case study of the Metropolitan Area of Milan
Giovanni Sanesi, Giuseppe Colangelo, Raffaele Lafortezza, Enrico Calvo and Clive Davies
Landscape Research, volume 42, issue 2 (February 2017) pp. 164–175

Chapter 4
Can we face the challenge: how to implement a theoretical concept of green infrastructure into planning practice? Warsaw case study
Barbara Szulczewska, Renata Giedych and Gabriela Maksymiuk
Landscape Research, volume 42, issue 2 (February 2017) pp. 176–194

Chapter 5
Siting green stormwater infrastructure in a neighbourhood to maximise secondary benefits: lessons learned from a pilot project
Danielle Dagenais, Isabelle Thomas and Sylvain Paquette
Landscape Research, volume 42, issue 2 (February 2017) pp. 195–210

For any permission-related enquiries please visit:
http://www.tandfonline.com/page/help/permissions

Notes on Contributors

Mariagrazia Agrimi is a Researcher at the Department for Innovation in Biological, Agro-food and Forest systems (DIBAF), University of Tuscia, Italy.

Anna Barbati is a Researcher at the Department for Innovation in Biological, Agro-food and Forest systems (DIBAF), University of Tuscia, Italy.

Enrico Calvo is a Project Leader at Regione Lombardia-ERSAF, Italy.

Giuseppe Colangelo is the Scientific Board Coordinator for Regione Lombardia-ERSAF, Italy.

Marcus Collier is a Senior Research Fellow at University College Dublin, Ireland, and the coordinator of the EU-FP7 TURAS project. His research interests focus on the interface between social and ecological systems, and his PhD research concentrated on damaged landscapes, resilience planning and collaborative processes.

Piermaria Corona is Director of the Forestry Research Centre of the Council for Agricultural Research and Economics (CRA), Italy, and Full Professor at the University of Tuscia, Italy.

Danielle Dagenais is Associate Professor in Landscape and the Environment of the Université de Montréal, Canada. She specialises in the art of gardening and ecology, plant identification and plant engineering, and horticulture. She has a particular focus on biodiversity, person–plant relationships, the ecological discourse of landscape architecture and the concept of urban nature.

Clive Davies is a Visiting Fellow at the School of Architecture, Planning and Landscape, Newcastle University, UK. His principle research interests are in green infrastructure planning, landscape planning, community governance and urban forestry.

Karen Foley is a Lecturer at the School of Architecture, University College Dublin, Ireland. Her research interests centre on urban open space, identifying tools and techniques to develop robust multifunctional landscape typologies in cities that satisfy social and environmental needs. Specifically, she is working in the area of green infrastructure, urban forestry and transitory open space.

Lorenza Gasparella is a doctoral student in Planning and Environmental Management and Landscape at the University of Rome, Italy.

Renata Giedych is Assistant Professor at the Department of Landscape Architecture, Warsaw University of Life Sciences, Poland. Her main field of research covers sustainable urban planning, urban green infrastructure, nature conservation in urban areas and legal basis for landscape planning.

NOTES ON CONTRIBUTORS

Gemma Jerome is a Green Infrastructure Project Manager at Gloucestershire Wildlife Trust. She is a first-class planning graduate from the University of Liverpool, UK, Gemma is working towards becoming a Chartered Town Planner.

Raffaele Lafortezza is a Professor in the Department of Agro-environmental and Territorial Sciences at University of Bari, Italy.

Mick Lennon is a Lecturer in Planning and Environmental Policy in the School of Architecture, Planning and Environmental Policy at University College Dublin, Ireland. His research interests centre on interpretative analysis, in particular how the meaning-making process influences approaches to environmental policy and planning.

Gabriela Maksymiuk is Assistant Professor at the Department of Landscape Architecture, Warsaw University of Life Sciences, Poland. Her research looks at instruments and drivers of urban green areas development, management of urban green areas and Green Infrastructure concept and possibilities of its implementation in Poland.

Ian C. Mell is a Lecturer in Environmental and Landscape Planning at University of Manchester, UK. He researches human–environment interactions exploring social, economic and ecological valuation of Green Infrastructure in the UK, EU, USA and Asia. He is the author of *Global Green Infrastructure: Lessons for Successful Policy-Making, Investment and Management* (2016, Routledge).

Sylvain Paquette is Associate Professor in Landscape and the Environment of the Université de Montréal, Canada. He specialises in the sociology of landscape, landscape management and spatial planning. He is interested in the social and cultural valorisation of inhabited territories, in questions of identity and quality of life (urban and peri-urban) and in the notion of new rurality.

Luigi Portoghesi is Associate Professor at the Department for Innovation in Biological, Agro-food and Forest systems (DIBAF), University of Tuscia, Italy.

Giovanni Sanesi is a Professor at the University of Bari, Italy. His research centres on Forestry, Urban Forestry, Timber Production and Wild Fires.

Mark Scott is Professor of Planning in the School of Architecture, Planning and Environmental Policy at University College Dublin, Ireland. His research interests are in the broad areas of environmental planning and land-use governance.

Barbara Szulczewska is Professor at the Department of Landscape Architecture, Warsaw University of Life Sciences, Poland. Her research interests include spatial problems of sustainable development, green open spaces development and landscape planning.

Isabelle Thomas is Full Professor at the Faculty of Planning, Université de Montréal, Canada. She specializes in issues of sustainable development, urban vulnerability and regional development. Her research also focuses on smart urban growth with an international perspective, and the use of geographic information systems to promote sustainable development.

Antonio Tomao is a Research Fellow at the Department for Innovation in Biological, Agro-food and Forest systems (DIBAF), University of Tuscia, Italy.

Alexander Whitehouse is a Development Planner at Hallam Land Management, UK.

Green infrastructure: reflections on past, present and future praxis

Introduction

When Lowenthal (1985) talked about the past being a foreign country, he discussed how our perceptions of the landscape were shaped by experience, by socio-cultural actions and by our changing commitment to ideas or causes. Similarly, when Benedict and McMahon (2006) emphasised that Green Infrastructure (GI) was, and is, our life-support system, they were reflecting on the growth of sustainability politics and a subsequent rethinking within the environment sector of how best to manage human–environmental interactions. Both Lowenthal and Benedict and McMahon explored the actions of the past to understand those currently visible in landscape planning. They also identified significant problems in how we deal with the landscapes around us and the impact this has on our ability to manage landscape resources sustainably. Neither offered a fool-proof answer to the questions of how we address this growing debate; Benedict and McMahon did, however, synthesise the ideas of parkways, greenways, environmental management and ecological conservation into a more holistic approach to landscape planning: what we now consider as a 'GI' approach.

This special issue of *Landscape Research* takes this debate as a starting point to examine how planners, academics, practitioners and other stakeholders are utilising GI principles. The special issue call asked the growing GI community to examine whether it was a sufficiently developed concept, what best practice could be identified, and to explain the nuances embedded within its implementation in different locations. Taking an overtly broad approach, the special issue provided scope for papers to address a range of thematic, spatial, innovative and conceptual understandings of GI in praxis. We feel that the papers presented in this special issue successfully achieve this by exploring contemporary understandings of the financial, ecological, policy-practice and scalar uses of GI. This does, however, raise a dilemma that is explored through the research articles and doctoral position papers presented. How has GI developed to meet these challenges? Within this debate, a series of key ideas are presented, which have been established in the academic literature as shaping the ways in which we *discuss, value,*[1] and *utilise* GI in alterative geo-political landscapes. The introduction draws together the historical discussions of GI to contextualise the papers presented in this special issue, which go on to explore the versatility of GI as an approach to landscape planning, and as a concept which can, as Benedict and McMahon remark, act as a life support system for human and ecological activities.

Throughout this editorial, and the subsequent papers, the meaning of GI will be broadly framed by the definitions proposed by Benedict and McMahon (2006), Natural England (2009) and more recently the European Commission (2013). Each of these definitions utilises a range of socio-economic and ecological principles, landscape resources, and alternative approaches to landscape planning to frame what GI *is, how* it should be developed and *what* benefits it should deliver. The key principles within this process are the promotion of social, economic and environmental benefits within an integrated approach to planning that enables different stakeholders to shape the ways that they develop and manage the landscape. Furthermore, the principles of multi-functionality, connectivity and access to nature, supportive ecological networks, and establishing socio-economic values through awareness raising and stewardship are all presented as essential components of the promotion of GI praxis

(Countryside Agency & Groundwork, 2005; Davies, Macfarlane, McGloin, & Roe, 2006; Mell, 2010; Weber, Sloan, & Wolf, 2006; Williamson, 2003).

The timing of the special issue is also apt. It comes almost a decade since GI started to be actively discussed by academics and practitioners[2] in the UK and USA; the initial discussions of Benedict and McMahon (2006) and Davies et al. (2006) being important milestones. These initial explorations led to the development of a burgeoning literature focussed on the impacts as well as the functionality of GI across the globe. A second reason why this issue of *Landscape Research* is timely is the growing realisation in government that alternative solutions to climate change and urban expansion are needed if we are to plan our landscapes more sustainably. Such a reposition is becoming increasing visible in the UK, North America and Europe, but also in the expanding economies of the BRICS countries: Brazil, China, India, Russia and South Africa (Moraes Victor et al., 2004; Schäffler & Swilling, 2012). Landscape planning and management therefore needs to be considered as a global discussion, and one where GI offers a suite of options that can, and have been evidenced to, mitigate climate change, alleviate flood risk, improve public health, and promote economic viability (Gill, Handley, Ennos, & Pauleit, 2007; Mell, Henneberry, Hehl-Lange, & Keskin, 2013; Ulrich, 1984; Weber et al., 2006). The same literature also illustrates that GI can be delivered successfully at a number of scales, and across a range of administrative and ecological boundaries (Davies et al., 2006; Forman, 1995; Kambites & Owen, 2006; Thomas & Littlewood, 2010). This evidence places our understanding and use of GI in a far more established position than it was for those who first strove to develop the concept in the early 2000s.

Over the course of the last decade, there has been year-on-year growth in the discussion and publication of GI research, guidance and policy, of which the authors of several papers in this special issue of *Landscape Research* have contributed. Research into GI has been varied, and in a sense, highlights the inherent versatility of the concept, which has both supported and hindered its uptake in different locations. The research of the Conservation Fund (Benedict & McMahon, 2006; Lerner & Allen, 2012; Weber & Wolf, 2000) in the USA, and the Community Forest Partnerships (Blackman & Thackray, 2007; Davies et al., 2006) and Natural England (2009) in the UK, and more recently regional landscape administrations in North-West Europe (South Yorkshire Forest Partnership & Sheffield City Council, 2012; Vandermeulen, Verspecht, Vermeire, Van Huylenbroeck, & Gellynck, 2011; Wilker & Rusche, 2013), have all played a significant role in ensuring that positive GI messages are visible within academic/practice debates.

Over the last five years, we have also seen governments engage with these messages: the US Environmental Protection Agency (EPA) have issued memorandums on GI, whilst the UK government mentions GI in the National Planning Policy Framework (NPPF) (Benedict & McMahon, 2006; Department of Communties & Local Government, 2012; EPA, 2014). More recently, the European Commission issued a communique supporting GI (European Commission, 2013), whilst researchers working for the New Zealand government presented a synthesis of GI and its proposed values for planning in that country (Boyle et al., 2013).

Such geographical diversity has enabled GI to develop as a dynamic approach to landscape planning; a process which is reflected in the scope of the discussions presented in this special issue. It has, however, also highlighted the complexity witnessed in many cities and regions in Europe, the USA, and more recently in African (Abbott, 2012) and Asian nations (Merk, Saussier, Staropoli, Slack, & Kim, 2012), where the use of GI has varied in its form and uptake. Variation has been linked directly to the level of understanding and support (political, social and financial) that GI receives from the government and the environmental sector (Siemens, 2011). Whilst this suggests that the uptake of GI at a government level is positive, it has been the promotion of a number of its key principles: multi-functionality, connectivity, ecological networks and integrated approaches to policy-implementation, by a select group of agencies that have driven the transition from policy to practice, for example, England's Community Forests (cf. Blackman & Thackray, 2007). Such diversity implies that a plurality to GI discussions exists which can be considered to be spatially, temporally and socio-politically constructed (Mell, 2014); dimensions which will be explored within the papers presented in this special issue of *Landscape Research*.

The three eras of GI: exploration, expansion and consolidation

To frame such a debate, it is necessary, as Lowenthal stated, to understand the past, in this case the historical development of GI as a concept and as an approach to planning. The versatility of GI lies in its synergies between a series of established green space antecedents such as greenways (Little, 1990), the Garden Cities movement (Howard, 2009) and landscape ecology (Jongman & Pungetti, 2004), the key principles of the concept, and the developing consensus between academics, practitioners, delivery agents and policy-makers, each of which has positioned GI as an integrated and cost-effective approach to urban and landscape planning. Although there is a clear line of argument, as discussed by Mell (2010), illustrating how these factors have generated an overarching acknowledgement of *what GI is, what it should do, and how it should do it*, we can go further and identify three periods of GI development: Exploration (1998[3]–2008), Expansion (post-2008–2011) and Consolidation (2010–2012 onwards).

GI developed through a process of assimilation and adaption. Its growth utilised the principles outlined in Ebenezer Howard's Garden Cities project to embed the notion of connectivity, accessibility and integrated planning at the core of GI praxis (Town & Country Planning Association, 2012a). GI has also embraced the notions of linearity and connectivity across and between landscape (and administrative) boundaries central to greenways planning to promote multi-functional ecological and recreational routes (Ahern, 1995; Fábos, 2004; Little, 1990). Moreover, GI has been influenced by complementary principles identified in the landscape ecology and conservation research (Farina, 2006; Forman, 1995; Weber & Wolf, 2000). Linking these issues is a view that GI should act as a holistic approach that integrates the socio-political and environmental concerns of landscape planning (Dunn, 2007; Natural England & Landuse Consultants, 2009; Young & McPherson, 2013).

The current discussions of GI engage with each of these issues, highlighting how it can be used to address a variety of planning mandates. Furthermore, given the dynamic nature of landscape planning, GI advocates are continuing to evidence the concept to address climatic and demographic change (Goode, 2006), *grey/built environment* vs. GI (Mell, 2013) debates, and the growing discussion of ecosystem services (Andersson et al., 2014). However, throughout this process there remains a need to understand how discussions of contemporary GI investment are linked to the history of green space planning.

Exploration (1998–2007)

The initial exploration of GI extended Parris Glendening's call for it to be used to optimise the ways in which we develop, plan and manage our landscapes (President's Council on Sustainable Development's, 1999). Although GI terminology was not initially used, this was what was being referred to through a range of synonyms, such as greenway planning or green space management. Benedict and McMahon (2002) published one of the first papers to consider GI using contemporary terminology. By linking its use to the smart conservation movement in North America, they acted as the catalysts for the expansion of debate surrounding GI. From this point onwards research and practitioner reports started to be populated with the term GI, using it to frame conservation discussions at a local and a regional scale (McDonald, Allen, Benedict, & O'Connor, 2005; Weber & Wolf, 2000). Over time, an acceptance of the terminology, and the ways in which it could be applied to landscape planning became increasingly visible (Benedict & McMahon, 2006).

The initial uptake of GI was seen to be steadily engaging European academics and practitioners. Evidence of this process was reported by Sandström (2002), who discussed the value of GI to Swedish planning debates, whilst Beatley (2000) debated comparable principles in his green urbanist assessment of a number of European cities. In the UK, England's Community Forest Partnerships (2004), and the Country In and Around Towns (CIAT) programmes of the Countryside Agency & Groundwork (2005) helped to focus GI discussions onto a number of key ideas: *connectivity, multi-functionality, interrelated* and *supportive* benefits, and a *systematic* (i.e. strategic) approach to landscape management. Each of these principles was used to shape how the environment sector and local government in England approached the revitalisation of landscapes across rural–urban boundaries. It also laid the ground work for further investigations into site specific applications of GI, based on the growing consensus of what it should deliver (Mell, 2010).

Each subsequent examination of GI extended these initial conceptualisations illustrating where, and how, it could be used as an effective form of landscape and urban planning. Key reference points in this process include Tzoulas et al.'s (2007) whose work reviews the potential health benefits of GI, work that drew on a wealth of further human-environmental research (Maas, Verheij, Groenewegen, de Vries, & Spreeuwenberg, 2006; Pretty et al., 2007; Pretty, Peacock, Sellens, & Griffin, 2005; Town & Country Planning Association, 2012b; Ulrich, 1984). Furthermore, Gill et al.'s (2007) and Goode's (2006) assessment of GI helped to synthesise its role in adapting of cities to climate change; research that has influenced more recent examinations of the utility of urban greenspace and green technology in a number of cities (cf. Bowler, Buyung-Ali, Knight, & Pullin, 2010; James et al., 2009). Ted Weber and the Conservation Fund's work in Maryland (USA) has also been used to frame contemporary ecosystem service debates, and remains highly influential in promoting links between investment in GI and the conservation of ecological networks (Weber, 2007; Weber et al., 2006; Weber & Wolf, 2000).

The work of each of these authors provided GI advocates with a number of avenues through which to explore the concept. They can also still be considered relevant to policy discussions, as they promote a more detailed appreciation of the varied socio-economic and ecological benefits of GI, and make significant contributions to the current thinking on landscape management. Extending this process, Kambites and Owen (2006) framed a number of these arguments within a discussion of how GI could more effectively engage planning policy debates. They offered one of the first, and potentially still most relevant assessments of what governance and policy structures GI needed to engage with if it was to become grounded in landscape praxis.

Expansion (2005–2010)

Following the initial period of exploration GI development moved into its expansion phase. This reflected an increase in the number of academic, government agencies and practitioners working with GI, and a significant increase in discussions, policy guidance and research projects looking at its benefits. In England, this was strongly linked to the inclusion of GI in the Regional Spatial Strategy (RSS) process, in which advocates used the growing body of research to relate its values to regional government (Horwood, 2011; Thomas & Littlewood, 2010). The RSS brought together a diverse range of partners to discuss the evolving concept, which was framed, to a large extent, by the work of the Community Forest network and their own landscape renewal programmes. However, the success of this process was variable with specific regions (the East of England, North-West, and the North-East), being the most proactive advocates of investment in GI (Blackman & Thackray, 2007; Llausàs & Roe, 2012; North West Green Infrastructure Think Tank, 2006).

Comparable praxis was also witnessed in the USA, where the continued influence of the Conservation Fund, and the release of Benedict and McMahon's seminal GI book (Benedict & McMahon, 2006), ensured the concept remained visible in landscape debates. One of the key factors contributing to this was the geo-political weight of the Conservation Fund, who have regional chapters, enabling them to initiate GI debates at a national, regional and sub-regional level. Throughout this period, they should be seen as the leading advocates for the development of GI in the USA, both conceptually and spatially. More recently, as an emphasis on stormwater and water catchment dynamics has been mainstreamed within North American GI debates, we have witnessed an increasing engagement by the EPA. They have supported GI through a series of memoranda outlining how GI should be used to manage water resources (Amati & Taylor, 2010; Dunn, 2010; EPA, 2014). This reflects the historic engineered approach to investment in water management but they have also started to think more holistically about how soft or green engineering can be used to promote more sustainable forms of investment (Ahern, 2007; Mell, 2013). Such thinking has also been manifested in how city and sub-regional agencies, such as the Chicago

Metropolitan Water Reclamation Districts (MWRD), have re-evaluated their approaches to urban water management (Mell, 2016).

The second period of GI development also saw a growing regional development of the concept. The research and 'grey' literature[4] saw an elaboration of how GI was being planned through strategic documents, and where investments were taking place at a site, city and sub-regional scale (Mell, 2010). The number of GI strategies increased greatly in the UK at this point, with a similar growth process being witnessed across the USA, where major cities including New York, Chicago and Philadelphia started to explore the possibilities of investing in GI as an effective approach to land management (Mell, 2014). Instigated in parallel to these discussions were examples of Spanish, Scandinavian and Italian explorations of GI utility which addressed climatic, functionality and administrative differences in its use across Europe (James et al., 2009; Lafortezza, Carrus, Sanesi, & Davies, 2009; Llausàs & Roe, 2012). This period therefore witnessed the beginnings of a more refined approach to GI that examined its value as a planning process and was framed using more nuanced interpretations. This differed from the initial stage, as there was a greater level of consensus amongst GI advocates regarding its guiding principles (Beer, 2010; Mell, 2008, 2010). This shift would gain momentum in the third phase: consolidation, currently taking place in GI research.

Consolidation (2010 onwards)

The current phase of GI research can be considered one of consolidation, in the sense that we have now established a relatively common consensus relating to what GI is and how it should be developed (Mell, 2013, 2014; Wright, 2011). This has been supported by the year-on-year growth of specific GI investment strategies, as well as an increasing awareness of its value to a number of thematic planning agendas (i.e. climate change). Therefore, whilst GI in the expansion period looked to assess *how*, *where* and *why* it could be seen as a relevant form of investment, the transition towards consolidation aimed to develop a more detailed, grounded and robust evidence base to support its use.

One factor influencing this process has been the growing realisation, especially within global cities, of the economic, ecological and social value that GI can help deliver (Siemens, 2011). Beatley's (2000, 2009) research in Europe and Australasia illustrates this broadening trend, as do the consolidation and adoption of GI strategies in the cities of Chicago (Chicago Metropolitan Agency for Planning, 2014) and London (Greater London Authority, 2012). Further guidance has also been produced by Merk et al. (2012) and Siemens (2011) reflecting on the opportunities for GI investment in Asia and examining the added-value of greener and more sustainable cities. What each of these strategies and guidance documents argued was that GI is now seen as a more appropriate approach to the delivery of multi-functional landscapes compared to other forms of development. GI is thus being linked to greener, smarter and more efficient methods of urban development (Austin, 2014; Hansen & Pauleit, 2014; Jones & Somper, 2014).

Furthermore, as city and national/international guidance has developed to frame GI investment, there has been a corresponding refinement in more specific thematic applications of the concept (European Commission, 2013). Austin (2014) discussed this process by assessing the potential applications of biodiversity and ecosystems in North America and Sweden to examine how climatic variation could be managed with GI inspired investment in green urban design and ecological corridors. Moreover, Rouse and Bunster-Ossa (2013) addressed the continuing reflection on grey vs. green water sensitive management in the USA. However, whilst both debated on where and how GI could be implemented these discussions have been supplemented by more detailed studies of how the concept can address geographically specific research focussed on financial concerns in North-West Europe (Mell et al., 2013; Vandermeulen et al., 2011; Wilker & Rusche, 2013); conservation and ecosystem services benefits in urban and suburban areas in the USA (Hostetler, Allen, & Meurk, 2011; Young, 2010). Plus the need to develop an evidence base for politicians emphasising the role of GI in adapting urban areas to climate change (Ahern, 2013; Carter & Fowler, 2008; Madureira, Andresen, & Monteiro, 2011).

One further aspect of GI thinking that remains crucial to our understanding of its function is that regardless of what form of investment GI takes, we must consider it to be context specific. Although the big questions of: *who, what, when,* and *how* may change, it remains important to reflect on the *where* and *why*, and the socio-economic, political and landscape factors that influence development. Furthermore, although we can identify a consensus of which GI principles are considered to be accepted in each of the major GI planning arenas, there is still scope to celebrate the differences between locations (Mell, 2016). Where variation in the use of GI exists, for example, between the UK and the USA (Mell, 2014), we can identify complementary and contrasting implementation that can facilitate increased reflections on best-practice between locations. Given the development challenges facing landscape and urban planners in China and India, there are ecological and economic benefits to understanding such transferability. What we do with GI, as well as how we evaluate these different approaches to investment therefore provides us with a more nuanced appreciation of its value in landscape planning discussions.

The GI special issue

The following set of papers explore a number of issues presented in the discussion above illustrating how GI has developed, and how it is currently being utilised in a number of different geo-political contexts. Each of the papers discusses the subtleties of GI research, making links between various physical, socio-economic and legislative-administrative factors. As a collection, the papers present a snap-shot of the discursive breadth of GI planning, a view which is often overlooked as GI research is prone to focusing on the specificities of individual projects. This special issue of *Landscape Research* thus adds depth to the evidence base for GI, enabling further examination of the context, content and outcomes of existing GI discussions. Exploring the versatility of GI enables us as commentators, researchers and practitioners to extend these debates into wider planning discussions, as well as extending the evidence base for those who support it.

The papers can be identified as integrating a number of the ideas noted in the three eras presented above, illustrating the growing understanding of GI characteristics, spatial application and thematic focus (cf. Mell, 2015). All seven papers could be considered to highlight the growing knowledge base within the field of GI, as they review the application of specific ecological, water-centric and socio-economic approaches to landscape planning in unique, yet comparable, contexts. The papers thus move beyond a simplistic categorisation of spatial or thematic approaches to GI planning. Alternatively, they report on the increasing use of best praxis to examine the interplay of geo-spatial understandings of landscape investment within the context specific political structures of planning in different locations. The papers also reflect on the interactivity of political, social and ecological systems to examine the complexity of achieving balance between economic valuation, maintaining ecological capacity, and the delivery of appropriate GI investment across urban, urban-fringe and rural/coastal landscapes (Austin, 2014; Rouse & Bunster-Ossa, 2013). Within each paper, this dynamic approach to GI is considered in terms of addressing contextually specific issues, i.e. stormwater management in Canada or how GI is being incorporated into the development process in Ireland, the latter including a reflection on the expanding understanding of how governance and consultation influences understanding of the investment process, and provides new interpretations of the existing principles of GI.

Lennon et al.'s paper focuses on the evolving discussions of GI as an approach to landscape planning in Ireland (Lennon, Scott, Collier, & Foley, 2016). Their paper evaluates the difficulty that planners have encountered in establishing a 'landscape perspective' both within the consultation process and in policy at different scales because of existing stakeholder mandates. GI is therefore proposed by Lennon et al. as an alternative approach which can draw on the multi-disciplinary expertise of engineers, planners and ecologists to identify innovative solutions to landscape planning. Moreover, they go on to assess how disciplinary silos are being eroded through an integrated approach to GI planning. However, although the policy landscape at a local and regional level appears to be engaged with the GI agenda, there is reluctance, as in many other countries such as the UK (Mell, 2016), to mandate GI within national level policy.

Sanesi et al.'s paper uses the urban forests of Milan, Italy as the central focus of their examination of GI (Sanesi, Colangelo, Lafortezza, Calvo, & Davies, 2016). Using a historical analysis of the city's woodland areas, Sanesi et al. discuss how urban woodlands have been promoted as a reaction to increased urbanisation from the 1970's onwards as an environmental defence mechanism to the loss of urban green space. They argue that GI provides a more nuanced understanding of the connections and fragmentations of urban systems compared with more traditional forms of green space planning. In particular, the paper traces the growing use of urban forest by Milan's city government as a successful method of providing accessible and functional spaces across the city.

The following paper by Szulczewska et al. discusses the development of GI praxis in Warsaw, Poland (Szulczewska, Giedych, & Maksymiuk, 2016). They examine the development of the multi-functionality focussed Warsaw Green Infrastructure (WGI) approach to investment which moves away from the historical (and inflexible) form of green space management, the Warsaw Natural System (WNS). The application of the WNS used landuse classifications to identify the spatial pattern of Warsaw but lacked further analysis of the functions associated with these spaces. The paper discusses the shift to the WGI model and how the city and its agencies, i.e. Environmental Protection Officers, identify the city's GI resource base and attempt to clarify how they deliver the socio-economic and ecological benefits associated with it across the city. The paper reports that a joint approach to investment is being used that aims to plan strategically through government policy/initiatives and locally by engaging with non-governmental officers and other agents such as Warsaw's Architecture and Urban Planning officers, to ensure that an appropriate network of GI is developed.

In contrast to the strategic policy and practice discussions of Sanesi et al. and Szulczewska et al., Dagenais et al.'s paper focuses on the role of green stormwater infrastructure (GSI), in the effective management of urban water systems in Beauport, Quebec City, Canada (Dagenais, Thomas, & Paquette, 2016). The paper explores the use of GSI as a mechanism for environmental, aesthetic and social benefits to both the physical landscape and the population of Beauport. This includes the piloting of the GSI approach in different sites and the potential to scale-up the benefits to a broader geographical region. It also engages with the complexity of consensus building between stakeholders and the difficulties faced by GI practitioners in promoting the benefits of investing in GI to a variety of different partners.

Moving from water to ecology, Gasparella et al.'s paper uses the management of Italian Stone Pines (*Pinus pinea* L.) in the urbanisation of the greater Rome region as the focus of its GI discussion (Gasparella et al., 2016). Reflecting on landuse change from 1949–2008, the authors use a synthetic index of landscape to outline how the increased use of Italian Stone Pines as a socio-economic amenity has brought it into direct contact with the challenges of managing urban growth. The authors note that planners and developers need to be aware of the complex regulatory framework controlling planning policy and environmental regulations if they are to retain the long-term ecological capacity of the forests. Perhaps, predictably, the paper finds that those areas with greater protection have seen limited change to the size or function of the forests, whilst other areas with less protection have witnessed greater change. Gasparella et al.'s paper does, however, provide historical evidence of the impacts that urbanisation has on forest cover and reflects on a range of protective methods, including buffer zones, which can be employed to ensure the ecological capacity of the forests is not undermined.

In addition to the discussions of praxis, the special issue provides space for two developing voices to enter GI planning debates to extend the existing interpretations of the concept noted previously. Two papers from current/recent doctoral students are presented providing scope for fresh analysis to be made on how we engage different communities at varying scales with GI, and how alternative valuation mechanisms can influence how we value the landscapes around us. Jerome's paper discusses the role that GI takes when used at a community-scale (Jerome, 2016), examining the complexity of utilising GI where stakeholder knowledge of the concept's terminology may limit the understanding of the socio-economic and ecological benefits that can be delivered at a neighbourhood or even a street-scale. Drawing on theoretical debates, Jerome links the diverse drivers of engagement with local greenspaces with the wider conceptions of participation to discuss the role of people, place, and context in the successful delivery of GI projects. Where Jerome discusses how GI is being *used* by community groups to support investment at the micro-scale, Whitehouse presents an analysis asking whether existing GI valuation toolkits make a positive contribution to *how* we value green spaces (Whitehouse, 2016). Whitehouse's work on North-West England, but framed in the wider global valuation debates, highlights the inherent complexity GI professionals face in valuing nature. This reflects on the wider conceptual discussions of 'value' but focuses on the role that valuation toolkits currently hold in establishing economic values for GI that are acceptable to planners, developers and environmental practitioners.

Both papers examine elements of GI, i.e. more effective engagement, project focus, economic value, which are embedded within wider GI research. They both illustrate, as do the five previous papers that understanding GI is not a simple process. On the contrary, it should be considered as a deeply complex interaction of people, place and politics, where the interactivity of policy and practice are central to the successful delivery of urban and landscape planning needs. The value of these discussions is therefore in highlighting the benefits associated with the value of GI to people outside of the active research-investment community.

Concluding remarks

GI has evolved extensively over the last ten years and has developed from an interesting extension of a number of existing green space planning activities into a defined and flexible approach to landscape planning. What we have witnessed is how GI has become established, along with the more recent promotion of ecosystem services, as one of the most engaging forms of landscape investment and management for planners. Due to its versatility, coupled with an integrated approach to policy-practice debates, GI thinking has set itself apart from other forms of planning (Mell, 2013). The papers in this special issue reflect this dynamism, illustrating how planning at a number of scales (neighbourhood, city and regional) with a range of delivery objectives (water management, urban forestry and community development) can lead to more responsive, and in many cases appropriate form of investment. In spite of such positivity, there is still scope to embed GI further within policy and practice. Planning policy in the UK remains varied in its support for GI, whilst in the USA the water-centric focus of GI still persists. Furthermore, in those locations that are starting out on their GI journey, e.g. India, Bangladesh, China, and South Africa, there is a small, yet growing evidence base of research focussing on socio-ecological connectivity that may develop into a normative form of urban planning. In summary the *how*, *why* and *what* we do in landscape and urban planning is changing, for, in my opinion, the better. However, we must not be reticent in our continual exploration of the evidence base and rationale for investment in GI. We must also continue to engage the public to ensure, as Louv (2005) discusses, that a long-term link between the landscape and people is maintained. Most of all we must continue to illustrate the value of GI to partners, advocates and politicians to ensure that it continues to form part of the discussion of planning delivery mechanisms.

Notes

1. Value in this sense reflects the collective ecological, economic, political and social value presented in the wider GI literature.
2. Although this Editorial recognises that the debates surrounding GI have been visible since the late 1990s, and were based on the antecedents of the greenway, landscape ecology, Garden Cities and sustainable urbanism literature, see Mell (2010) for further details.
3. One of the first reported uses of 'GI' terminology was by the former governor of Maryland (USA), Parris Glendening, who discussed its value within the President's Council on Sustainable Development (1999). In the USA GI became increasingly prominent following Benedict and McMahon's (2002) discussion of the concept, with the UK developing its initial discussion from 2004 onwards (Mell, 2010).
4. The 'grey' literature is considered as the practitioner and policy literature developed by government and advocacy organisations. This research is often presented in the form of policy and guidance documents and is less likely to appear to within the academic literature.

Funding

This work was supported by the U.S. Department of Housing and Urban Development.

References

Abbott, J. (2012). *Green infrastructure for sustainable urban development in Africa*. London: Routledge.

Ahern, J. (1995). Greenways as a planning strategy. *Landscape and Urban Planning, 33*, 131–155.

Ahern, J. (2007). Planning and design for sustainable and resilient cities: Theories, strategies and best practice for green infrastructure. In V. Novotny, J. Ahern, & P. Brown (Eds.), *Water-centric sustainable communities* (pp. 135–176). Hoboken, NJ: Wiley-Blackwell.

Ahern, J. (2013). Urban landscape sustainability and resilience: The promise and challenges of integrating ecology with urban planning and design. *Landscape Ecology, 28*, 1203–1212.

Amati, M., & Taylor, L. (2010). From green belts to green infrastructure. *Planning Practice and Research, 25*, 143–155.

Andersson, E., Barthel, S., Borgström, S., Colding, J., Elmqvist, T., Folke, C., & Gren, A. (2014). Reconnecting cities to the biosphere: Stewardship of green infrastructure and urban ecosystem services. *Ambio, 43*, 445–453.

Austin, G. (2014). *Green infrastructure for landscape planning: Integrating human and natural systems*. New York, NY: Routledge.

Beatley, T. (2000). *Green urbanism: Learning from European cities*. Washington, DC: Island Press.

Beatley, T. (2009). *Green urbanism down under: Learning from sustainable communities in Australia*. Washington, DC: Island Press.

Beer, A. R. (2010). Greenspaces, green structure, and green infrastructure planning. In J. Aitkenhead-Peterson & A. Volder (Eds.), *Urban ecosystem ecology* (pp. 431–448). Madison, WI: American Society of Agronomy Inc, Crop Science Society of America Inc, Soil Science Society of America Inc.

Benedict, M. A., & McMahon, E. T. (2002). Green infrastructure: Smart conservation for the 21st century. *Renewable Resources Journal*, 12–17. Autumn Edi.

Benedict, M. A., & McMahon, E. T. (2006). *Green infrastructure: Linking landscapes and communities. Urban Land* (Vol. June). Washington, DC: Island Press.

Blackman, D., & Thackray, R. (2007). *The green infrastructure of sustainable communities*. North Allerton: England's Community Forests Partnership.

Bowler, D. E., Buyung-Ali, L., Knight, T. M., & Pullin, A. S. (2010). Urban greening to cool towns and cities: A systematic review of the empirical evidence. *Landscape and Urban Planning, 97*, 147–155. doi:10.1016/j.landurbplan.2010.05.006

Boyle, C., Gamage, G., Burns, B., Fassman, E., Knight-Lenihan, S., Schwendenmann, L., & Thresher, W. (2013). *Greening cities: A review of green infrastructure*. Auckland: University of Auckland.

Carter, T., & Fowler, L. (2008). Establishing green roof infrastructure through environmental policy instruments. *Environmental Management, 42*, 151–164.

Chicago Metropolitan Agency for Planning. (2014). *GOTO 2040 comprehensive regional plan*. Chicago, IL: Author.

Countryside Agency & Groundwork. (2005). *The Countryside in and around towns: A vision for connecting town and county in the pursuit of sustainable development*. Wetherby: Countryside Agency.

Dagenais, D., Thomas, I., & Paquette, S. (2016). Siting green stormwater infrastructure in a neighbourhood to maximise secondary benefits: Lessons learned from a pilot project. *Landscape Research, 42*, 195–210. doi: 10.1080/01426397.2016.1228861.

Davies, C., Macfarlane, R., McGloin, C., & Roe, M. (2006). *Green infrastructure planning guide*. Anfield Plain: North East Community Forest.

Department of Communties and Local Government. (2012). *National planning policy framework*. London: Author.

Dunn, A. D. (2007, May–June). Green light for green infrastructure. *Pace Law Review*, pp. i–iv.

Dunn, A. D. (2010). Sitting green infrastructure: Legal and public solutions to alleviate urban poverty and promote healthy communities. *Boston College Environmental Affairs Law Review, 37*, 41–66.

England's Community Forests. (2004). *Quality of place, quality of life*. Newcastle: England's Community Forest Partnership.

Environmental Protection Agency. (2014). *Environmental protection agency*. Retrieved July 4, 2015, from http://www.epa.gov/

European Commission (2013). *Communication from the commission to the European parliament, the council, the European economic and social committee and the committee of the regions: Green Infrastructure (GI)—Enhancing Europe's natural capital*. Brussels: Author.

Fábos, J. G. (2004). Greenway planning in the United States: Its origins and recent case studies. *Landscape and Urban Planning, 68*, 321–342.

Farina, A. (2006). *Principles and methods in landscape ecology: Towards a science of the landscape*. London: Springer.

Forman, R. (1995). *Land mosaics: The ecology of landscapes and regions*. Cambridge: Cambridge University Press.

Gasparella, L., Tomao, A., Agrimi, M., Corona, P., Portoghesi, L., & Barbati, A. (2016). Italian stone pine forests under Rome's siege: Learning from the past to protect their future. *Landscape Research, 42*, 211–222. doi: 10.1080/01426397.2016.1228862.

Gill, S. E., Handley, J. F., Ennos, A. R., & Pauleit, S. (2007). Adapting cities for climate change: The role of the green infrastructure. *Built Environment, 33*, 115–133. Retrieved from https://doi.org/10.2148/benv.33.1.115

Goode, D. (2006). *Green infrastructure: Report to the Royal Commission on Environmental Pollution*. London: Royal Commission on Environmental Pollution.

Greater London Authority. (2012). *Green Infrastructure and Open Environments: The All London Green Grid. Supplementary Planning Guidance, London Plan 2011 Implementation Framework*. London: Author.

Hansen, R., & Pauleit, S. (2014). From multifunctionality to multiple ecosystem services? A conceptual framework for multifunctionality in green infrastructure planning for urban areas. *Ambio, 43*, 516–529.

Horwood, K. (2011). Green infrastructure: Reconciling urban green space and regional economic development: Lessons learnt from experience in England's north-west region. *Local Environment: The International Journal of Justice and Sustainability, 16*, 37–41.

Hostetler, M., Allen, W., & Meurk, C. (2011). Conserving urban biodiversity? Creating green infrastructure is only the first step. *Landscape and Urban Planning, 100*, 369–371.

Howard, E. (2009). *Garden cities of to-morrow (Illustrated Edition)*. Gloucester: Dodo Press.

James, P., Tzoulas, K., Adams, M. D., Barber, A., Box, J., Breuste, J., ... Ward Thompson, C. (2009). Towards an integrated understanding of green space in the European built environment. *Urban Forestry & Urban Greening, 8*, 65–75. doi:10.1016/j.ufug.2009.02.001

Jerome, G. (2016). Defining community-scale green infrastructure. *Landscape Research, 42*, 223–229. doi: 10.1080/01426397.2016.1229463.

Jones, S., & Somper, C. (2014). The role of green infrastructure in climate change adaptation in London. *The Geographical Journal, 180*, 191–196.

Jongman, R., & Pungetti, G. (2004). In R. Jongman & G. Pungetti (Eds.), *Ecological networks and greenways* (pp. 7–33). Cambridge: Cambridge University Press.

Kambites, C., & Owen, S. (2006). Renewed prospects for green infrastructure planning in the UK. *Planning Practice and Research, 21*, 483–496.

Lafortezza, R., Carrus, G., Sanesi, G., & Davies, C. (2009). Benefits and well-being perceived by people visiting green spaces in periods of heat stress. *Urban Forestry & Urban Greening, 8*, 97–108. doi:10.1016/j.ufug.2009.02.003

Lennon, M., Scott, M., Collier, M., & Foley, K. (2016). The emergence of green infrastructure as promoting the centralisation of a landscape perspective in spatial planning - the case of Ireland. *Landscape Research, 42*, 146–163. doi: 10.1080/01426397.2016.1229460.

Lerner, J., & Allen, W. L. (2012). Landscape-scale green infrastructure investments as a climate adaptation strategy: A case example for the Midwest United States. *Environmental Practice, 14*, 45–56.

Little, C. (1990). *Greenways for America*. Baltimore, MD: The John Hopkins University Press.

Llausàs, A., & Roe, M. (2012). Green infrastructure planning: Cross-national analysis between the North East of England (UK) and Catalonia (Spain). *European Planning Studies, 20*, 641–663.

Louv, R. (2005). *Last child in the woods: Saving our children from nature-deficit disorder*. Chapel Hill, NC: Algonquin Books.

Lowenthal, D. (1985). *The past is a foreign country*. Cambridge: Cambridge University Press.

Maas, J., Verheij, R. A., Groenewegen, P. P., de Vries, S., & Spreeuwenberg, P. (2006). Green space, urbanity, and health: How strong is the relation? *Journal of Epidemiology and Community Health, 60*, 587–592.

Madureira, H., Andresen, T., & Monteiro, A. (2011). Green structure and planning evolution in Porto. *Urban Forestry & Urban Greening, 10*, 141–149.

McDonald, L., Allen, W., Benedict, M. A., & O'Connor, K. (2005). Green infrastructure plan evaluation frameworks. *Journal of Conservation Planning, 1*, 12–43.

Mell, I. C. (2008). Green Infrstructure: Concepts and planning. *FORUM – E-Journal, 8*, 69–80.

Mell, I. C. (2010). *Green infrastructure: Concepts, perceptions and its use in spatial planning*. Newcastle: University of Newcastle.

Mell, I. C. (2013). Can you tell a green field from a cold steel rail? Examining the "green" of green infrastructure development. *Local Environment: The International Journal of Justice and Sustainability: The International Journal of Justice and Sustainability, 18*, 37–41.

Mell, I. C. (2014). Aligning fragmented planning structures through a green infrastructure approach to urban development in the UK and USA. *Urban Forestry & Urban Greening, 13*, 612–620.

Mell, I. C. (2015). Green infrastructure planning: Policy and objectives. In D. Sinnett, S. Burgess, & N. Smith (Eds.), *Handbook on green infrastructure* (pp. 105–123). Cheltenham: Edward Elgar Publishing.

Mell, I. C. (2016). *Global Green infrastructure: Lessons for successful policy-making, investment and management*. Abingdon: Routledge.

Mell, I. C., Henneberry, J., Hehl-Lange, S., & Keskin, B. (2013). Promoting urban greening: Valuing the development of green infrastructure investments in the urban core of Manchester, UK. *Urban Forestry & Urban Greening, 12*, 296–306.

Merk, O., Saussier, S., Staropoli, C., Slack, E., & Kim, J.-H. (2012). *Financing green infrastructure: OECD Regional Development* (Working Papers 2012/10). London: OECD.

Moraes Victor, R. A. B., Costa Neto, J. de B., Nacib Ab'Saber, A., Serrano, O., Domingos, M., Pires, B. C. C., ... Moraes Victor, M. A. (2004). Application of the biosphere reserve concept to urban areas: The case of Sao Paulo City Green Belt Biosphere Reserve, Brazil-Sao Paulo Forest Institute: A case study for UNESCO. *Annals of the New York Academy of Sciences, 1023*, 237–281. doi:10.1196/annals.1319.012

Natural England & Landuse Consultants. (2009). *Green infrastructure guidance*. Peterborough: Author.

North West Green Infrastructure Think Tank. (2006). *North west green infrastructure guide*. Retrieved from http://www. greeninfrastructurenw.co.uk/resources/GIguide.pdf

President's Council on Sustainable Development. (1999). *Towards a sustainable America, advancing prosperity, opportunity and a healthy environment for the 21st Century*. Washington, DC: Author.

Pretty, J., Peacock, J., Sellens, M., & Griffin, M. (2005). The mental and physical health outcomes of green exercise. *International Journal of Environmental Health Research, 15*, 319–337.

Pretty, J., Peacock, J., Hine, R., Sellens, M., South, N., & Griffin, M. (2007). Green exercise in the UK countryside: Effects on health and psychological well-being, and implications for policy and planning. *Journal of Environmental Planning and Management, 50*, 211–231.

Rouse, D. C., & Bunster-Ossa, I. (2013). *Green infrastructure: A landscape approach* [Paperback]. Chicago, IL: APA Planners Press.

Sandström, U. (2002). Green infrastructure planning in urban Sweden. *Planning Practice and Research, 17*, 37–41.

Sanesi, G., Colangelo, G., Lafortezza, R., Calvo, E., & Davies, C. (2016). Urban green infrastructure and urban forests: A case study of the Metropolitan Area of Milan. *Landscape Research, 42*, 164–175. doi: 10.1080/01426397.2016.1173658.

Schäffler, A., & Swilling, M. (2012). Valuing green infrastructure in an urban environment under pressure—The Johannesburg case. *Ecological Economics, 86*, 246–257.

Siemens, A. G. (2011). *Asian Green City Index: Assessing the environmental performance of Asia's major cities*. Munich: Author.

South Yorkshire Forest Partnership & Sheffield City Council. (2012). *The VALUE project: The final report*. Sheffield.

Szulczewska, B., Giedych, R., & Maksymiuk, G. (2016). Can we face the challenge: How to implement a theoretical concept of green infrastructure into planning practice? Warsaw case study. *Landscape Research, 42*, 176–194. doi: 10.1080/01426397.2016.1240764.

Thomas, K., & Littlewood, S. (2010). From green belts to green infrastructure? The evolution of a new concept in the emerging soft governance of spatial strategies. *Planning Practice and Research, 25*, 203–222.

Town & Country Planning Association. (2012a). *Creating garden cities and suburbs today: Policies, practices, partnerships and model approaches—A report of the garden cities and suburbs expert group*. London: Author.

Town & Country Planning Association (2012b). *Reuniting health with planning—Healthier homes, healthier communities*. London: Author.

Tzoulas, K., Korpela, K., Venn, S., Yli-Pelkonen, V., Kaźmierczak, A., Niemela, J., & James, P. (2007). Promoting ecosystem and human health in urban areas using Green Infrastructure: A literature review. *Landscape and Urban Planning, 81*, 167–178.

Ulrich, R. S. (1984). View through a window may influence recovery from surgery. *Science, 224*, 420–421.

Vandermeulen, V., Verspecht, A., Vermeire, B., Van Huylenbroeck, G., & Gellynck, X. (2011). The use of economic valuation to create public support for green infrastructure investments in urban areas. *Landscape and Urban Planning, 103*, 198–206.

Weber, T. (2007). *Ecosystem services in Cecil County's green infrastructure: Technical report for the Cecil County green infrastructure plan*. Annapolis, MD.

Weber, T., & Wolf, J. (2000). Maryland's green infrastructure—Using landscape assessment tools to identify a regional conservation strategy. *Environmental Monitoring and Assessment, 63*, 265–277.

Weber, T., Sloan, A., & Wolf, J. (2006). Maryland's green infrastructure assessment: Development of a comprehensive approach to land conservation. *Landscape and Urban Planning, 77*, 94–110.

Whitehouse, A. (2016). Common economic oversights in green infrastructure valuation. *Landscape Research, 42*, 230–234. doi: 10.1080/01426397.2016.1228860.

Wilker, J., & Rusche, K. (2013). Economic valuation as a tool to support decision-making in strategic green infrastructure planning. *Local Environment, 19*, 702–713. doi:10.1080/13549839.2013.855181

Williamson, K. S. (2003). *Growing with green infrastructure*. Doylestown, PA: Heritage Conservancy.

Wright, H. (2011). Understanding green infrastructure: The development of a contested concept in England. *Local Environment: The International Journal of Justice and Sustainability, 16*, 37–41.

Young, R. F. (2010). Managing municipal green space for ecosystem services. *Urban Forestry & Urban Greening, 9*, 313–321.

Young, R. F., & McPherson, E. G. (2013). Governing metropolitan green infrastructure in the United States. *Landscape and Urban Planning, 109*, 67–75.

Ian C. Mell

The emergence of green infrastructure as promoting the centralisation of a landscape perspective in spatial planning—the case of Ireland

Mick Lennon, Mark Scott, Marcus Collier and Karen Foley

ABSTRACT

The 'landscape' approach to planning and design has long since advanced a social–ecological perspective that conceives ecosystems health and human well-being as mutually constitutive. However, conventional public sector organisational arrangements segregate and discretely administer development issues, thereby militating against the holistic viewpoint necessary to redress the entwined nature of complex planning issues. The emergence and continuing evolution of green infrastructure (GI) thinking seeks to redress this problem by promoting interdisciplinary collaboration to deliver connected and functionally integrated environments. This paper reflects upon the ongoing development and institutionalisation of GI in Ireland as a means to critically evaluate 'if', 'why' and 'how' GI thinking promotes the centralisation of landscape principles in public sector planning. Drawing on a review of local authority practices and interviews with local authority officials, the paper traces and explains the concept's growth from the 'rebranding' of ecological networks to its current manifestation as a new mode of collaborative planning for multifunctional environments. This material is then employed to discuss the potential benefits and barriers encountered by GI planning more generally. Lessons are subsequently extrapolated for the advancement of landscape principles through innovative GI planning practices in other jurisdictions.

1. Introduction

Planning policy furnishes the framework for the future use of land. Therefore, it is inherently related to the fate of landscapes and the direction of landscape research. Consequently, a mainstay of activity for many of those engaged with the field of landscape research and practice has been the promotion of more holistic thinking in planning policy formulation to account for the complexities of social and ecological interactions (Ahern, Cilliers, & Niemelä, 2014; Benson & Roe, 2007; Selman, 2012). The emergence of 'social-ecological systems' thinking in spatial planning debates represents a recent turn in efforts to acknowledge this complexity and reorient thinking towards a more holistic perspective on the fundamental entwining of social and natural environments (Davoudi et al., 2012; Folke et al., 2010; Folke, Colding, & Berkes, 2003; Walker et al., 2006). As such, thinking in terms of social–ecological systems signifies the potential to centralise in planning policy those social–ecological relationships that have occupied much landscape research. Planning theorists in particular have seen promise in this perspective and have recently focused attention on locating ways to enhance the 'resilience' of such systems to a variety of environmental, political and institutional stressors (Wilkinson, 2012b). This has entailed a flurry of thinking on how the goals and objectives of planning can be adjusted to better account for social–ecological systems and how the resilience of such systems can be advanced (Cumming, 2011; Davoudi et al., 2012; Scott, 2013). Nevertheless, there remains a paucity of examples to illustrate what planning for social–ecological resilience might look like in practice and what forms of planning activity are required for its realisation (Wagenaar & Wilkinson, 2013). In essence therefore, there exists a lacuna in our understanding of how the holism of a landscape perspective may be effectively integrated into spatial planning practice.

This paper seeks to address this knowledge gap by reflecting upon the development and institutionalisation of the 'green infrastructure' (GI) approach in Ireland as a means to critically evaluate 'if' and 'how' it promotes the centralisation of a landscape perspective in planning practice. As such, this paper contributes to debates on substantive issues in landscape research concerning how planning activity should be conducted in a more self-reflective, responsive and holistic manner (Forester, 2013; Rydin, 2007). GI is an emerging and continually developing concept whose meaning is often dependent on who is employing it and the context in which it is deployed (Lennon & Scott, 2014). Use of the GI concept in Ireland is no different (Lennon, 2014). Consequently, this paper will trace the rise, evolution and institutionalisation of the GI concept in Ireland as a means of exploring its potential to position landscape concerns at the heart of planning practice. Ireland supplies an exceptionally good case study in which to trace the emergence, evolution and integration of this more holistic perspective in planning due to its particular administration and demographic attributes. Specifically, county and city development plans constitute the principal policy guidance document for land use planning at the local level in Ireland. These documents are produced under strictly prescribed timelines that require their review and adoption every six years. Giving more localised effect to the policies of these development plans are local area plans which are required to be reviewed every six years, subject to some dispensations.[1] As a result, it is feasible to trace the progression and transformation of a new planning policy concept throughout the comparatively frequent and recurring plan review process. Thus, the next section details the research methods adopted in gathering and analysing the empirical data used in this paper to trace the emergence and evolution of the GI concept in Ireland. The subsequent section discusses the theory of 'social-ecological resilience'. This is then employed to inform the scrutiny of the emergence of GI in Ireland conducted in the ensuing section. Following this, the paper presents an illustrative case study analysis of how a GI approach may give form to social–ecological resilience thinking in planning policy. The paper concludes by drawing lessons from the Irish experience on how a GI approach may help centralise a landscape perspective in spatial planning.

2. Research methods

This paper draws on the complementary and sequentially related research methods of documentary analysis and interviews. The documentary analysis entailed the scrutiny of 153 Irish policy documents identified as relevant to the study and assembled as an 'archive' (Foucault, 1972). This archive included plans, strategies and studies produced by a spectrum of national, regional and local governmental authorities, quasi-autonomous organisations and non-governmental organisations. The contents of the archive spanned the period from the first mention of GI in 2002 to November 2013 when it was considered that sufficient information had been collated and analysed to facilitate progression to the next stage of the research process. In particular, the examination of documentary material conducted enabled the confident determination of which planning authorities were leaders in advocating the GI approach. Two local planning authorities were identified, namely, Fingal County Council (FCC) and Dublin City Council (DCC). This procedure allowed the research team to locate a series of potential interviewees who it was considered beneficial to consult in seeking to understand the processes that facilitated the emergence, evolution and institutionalisation of the GI approach in each of the identified planning authorities.

A series of interviews were subsequently conducted between December 2013 and March 2014. A total of 17 people were interviewed. Fifteen of these were local authority officials and two were consultants who had recently worked closely with these authorities in formulating local area plans that promoted a GI approach through both land use policy and design specification. The interviewee selection process was based upon the level of involvement of the interviewees in the development of recent planning and design guidance that explicitly advanced the GI approach. This selection process was also grounded in a desire to represent a broad array of disciplinary perspectives in order to explore potential variations of opinion between different disciplines regarding the benefits of the GI approach. Those interviewed included policy and development management planners, ecologists, landscape architects, drainage and transportation engineers, a heritage officer,[2] urban designers and those in local authority management positions. The interview duration was on average 1 h 15 min. The interviews were conducted in a semi-structured format as this enabled 'openness to change of sequence and forms of questions in order to follow up the answers given and the stories told by the subjects' (Kvale, 1996, p. 124). Nevertheless, to ensure research consistency and that all issues relevant to the investigation were appropriately addressed (Bryman, 2008), the content of each interview was framed by a master interview guide that posed a series of 'essential questions' (Berg, 2004). Additional interviewee-specific questions were carefully tailored to reflect the particular position and potential insight of each interviewee.

This investigative process enabled the research team to establish that although both DCC and FCC invest much effort in promoting the GI concept in their respective planning activities, FCC is more advanced in progressing landscape scale social–ecological resilience. Consequently, in seeking to balance the constraints of space restrictions with a desire to ensure an adequate level of 'richness' (Geertz, 1973) in the analysis of data, this paper's detailed examination of local-level planning focuses upon the attributes and activities of FCC. Hence, drawing on material from nine of the interviews, the paper explores 'how' the officers of FCC have sought to overcome the limitations of traditional planning approaches by innovatively employing the GI concept in developing policy and design ideas for the urban fringe of Dublin City. This is undertaken by investigating the central processes and perspectives deployed to integrate a more holistic and contextually sensitive landscape perspective into spatial planning activities. However, to fully appreciate how this has been achieved, an understanding of social–ecological resilience is first required. Thus, the next section outlines the central tenets of social–ecological resilience and reviews debates surrounding the concept.

3. Social–ecological systems and resilience

Humanity is most often conceived as acting upon ecological systems rather than constituting an element of such systems (Coates, 1998; Goudie, 2009). Through this lens, management of ecological systems is seen to entail governance of a world external to, but influencing the well-being of society. However, since the early 1970s, there has emerged a growing awareness that human and ecological influence is profoundly interconnected and therefore inseparable (Folke, 2006). Now a perspective frequently evident across a range of disciplines, this view contends that many of the problems in natural resource management stem from a failure to acknowledge these inextricable connections (Folke et al., 2010). Thus, envisaging a world comprising complex and inter-linked 'social-ecological systems' is thought to better reflect human–environment relations. In this sense, humanity is conceived as a constituent in a system with compound interdependent feedback loops that determine the system's overall dynamics (Glaser, Krause, Ratter, & Welp, 2012). Accordingly, the concept reflects the principles grounding much landscape research by emphasising humans 'as' and 'in' nature rather than separate to and above nature (Ingold, 2000; Wylie, 2005). Furthermore, in keeping with the perspectives advanced by pioneers of the landscape approach such as McHarg (1969) and Spirn (1984), these social–ecological systems are understood to operate at multiple interrelated spatial and temporal scales. Each system is considered a semi-autonomous structure nested within a hierarchy of systems (Steiner, 2002, 2008). Hence, each system comprises a subsystem of another system in the hierarchy, and in turn, contains a number of subsystems within itself (Gunderson & Holling, 2001). The interactions across these system scales are thought fundamental in shaping the dynamics at any particular focal scale (Teigão dos Santos & Partidário, 2011). From this perspective, for example, a neighbourhood, municipal park, city, river catchment and state may all represent interrelated subsystem levels in a broader social–ecological system.

In recent years, research concerning social–ecological systems has increasingly been strongly associated with the concept of 'resilience' (Ahern, 2011, 2013; Collier et al., 2013; Pickett, Cadenasso, & Grove, 2004; Teigão dos Santos & Partidário, 2011). Thus, appreciating how landscapes may be influenced by planning's turn to this view of human–environment interactions necessitates attention to debates on the meaning and potential applications of 'resilience' thinking. Resilience is essentially a heuristic for thinking about change management. Fundamental to the concept is an assumption of non-linear dynamics in complex, nested and interrelated hierarchical systems (Eraydin & Taşan-Kok, 2012; Folke, 2006). The term emerged in the context of systems ecology where it was used to describe the ability of ecosystems 'to absorb changes of state variables, driving variables, and parameters, and still persist' (Holling, 1973, p. 17). Subsequent to its initial use, the expression has been employed across a range of disciplines from psychology (Norris, Stevens, Pfefferbaum, Wyche, & Pfefferbaum, 2008) and regional economic development (Dawley, Pike, & Tomaney, 2010; Pendall, Foster, & Cowell, 2010), to national security (Lentzos & Rose, 2009) and urban planning (Evans, 2011; Wilkinson, 2012b). However, it is its use within the ambit of social–ecological systems planning and management that primarily concerns this paper. Many of those employing the term seek to use it to help shift planning towards a more adaptable activity that is responsive to disturbance. In such instances, use of the concept in planning is assigned a normative content. In particular, those employing the term envisage that management for greater resilience opens up desirable pathways for development in a world where the future is difficult to predict (Barr & Devine-Wright, 2012; Plieninger & Bieling, 2012a).

Much contemporary debate concerning the use of resilience in planning centres on the distinction between 'equilibrium' and 'evolutionary' interpretations of the concept (Scott, 2013). The former understanding has its roots in disaster management and concerns a 'survival discourse' that focuses upon the ability of a system to 'bounce back' towards 'business as usual' following a catastrophe (Shaw & Maythorne, 2013). However, this perspective has received criticism concerning the appropriateness of seeking system persistence rather than adaptation when a crisis emerges (Davidson, 2010). In contrast to equilibrium-based approaches, 'evolutionary resilience rejects the notion of single-state equilibrium or a "return to normal", instead highlighting ongoing evolutionary change processes and emphasising adaptive behaviour' (Scott, 2013, p. 600). This interpretation focuses on resilience as enabling transformation of social–ecological dynamics such that disturbance supplies the stimulus for re-invention and thereby ensures strength through continuing reflection and adaptability (Erixon, Borgström, & Andersson, 2013). Hence, an evolutionary interpretation of resilience entails a more radical and optimistic perspective that embraces the opportunity to 'bounce forward' (Shaw & Maythorne, 2013). It seeks to supplant a desire for stability with the acceptance of inevitable change such that it inverts conventional modes of thought by 'assuming change and explaining stability, instead of assuming stability and explaining change' (Folke et al., 2003, p. 352). Here, thinking in terms of resilience is thought to encourage flexible responses to the constraints of land use and landscape planning (Ahern, 2013; Erixon et al., 2013), adaptability to broader environmental and economic disturbance (Fünfgeld & McEvoy, 2012; Haider, Quinlan, & Peterson, 2012; Pike, Dawley, & Tomaney, 2010), and a capacity for positive institutional evolution (Scott, 2013; Shaw, 2012; Teigão dos Santos & Partidário, 2011). It is from such perspectives that the concept is seen to help inform human–nature interactions, most prominently through theorising about social–ecological resilience.

In this context, social–ecological resilience is a framing device that merges the concepts of 'social-ecological systems' with 'evolutionary resilience' to inform planning for human–nature relationships in changing contexts. In essence, it seeks to provide a means for considering 'how to innovate and transform into new more desirable configurations' (Folke, 2006, p. 260). Social–ecological resilience thus amalgamates a descriptive viewpoint with an analytic perspective and normative position. Accordingly, those advocating this approach see it as both a scientific discipline and a governance discourse

(Wilkinson, 2012a). Thinking on social–ecological resilience may thus be seen as displacing discourses of 'sustainable development'. Although Scott (2013, p. 601) notes how many authors conceive it 'as a means to further elaborate (rather than replace) sustainable development', there is a fundamental difference between traditional approaches to sustainable development as conceived in an Irish context and the more dynamic focus of social–ecological resilience. This centres on divergent perspectives regarding the process of transition towards a more sustainable future. For example, in its 'key principles', the national 'Planning Policy Statement' that sets the strategic framework for spatial planning in Ireland states that,

> Planning must proactively drive and support sustainable development, integrating consideration of its economic, social and environmental aspects at the earliest stage to deliver the homes, business and employment space, infrastructure and thriving urban and rural locations in an economically viable manner that will sustain recovery and our future prosperity. (Department of Environment, Community and Local Government, 2015, p. 2)

This interpretation of sustainable development focuses on locating an optimal development path and then pursuing such a course in advancing a knowable trajectory towards 'future prosperity'. Hence, this interpretation of sustainable development assumes an ability to predict and plan for a state of sustainability that is durable, stable and normalised. However, in keeping with contemporary debates in landscape research (Plieninger & Bieling, 2012b), enhancing the resilience of social–ecological systems involves a more holistic approach to embracing change that emphasises ongoing adaptation (Walker & Salt, 2006). It promotes continuous experimentation (Evans, 2011) and accommodates the trial of novel ideas (Ahern, 2011). Consequently, thinking in terms of social–ecological resilience presents a more dynamic perspective than conventional understandings of sustainable development in Irish planning by reconfiguring the basic principles guiding thought and action. GI can be understood as a way to give practice-based form to abstract theoretical concepts concerning social–ecological resilience. In doing so, the GI approach can be seen as a means of centralising in planning practice the holistic perspective of much landscape research that conceives ecosystems health and human well-being as inherently entwined and mutually constitutive. Addressing such challenges requires a sea-change in land use governance in terms of the more effective integration of the ecological dimension alongside traditional planning concerns, implying a shift in institutional and organisational arrangements to reflect interdisciplinary collaboration. In the next section, we chart the emergence and evolution of GI in spatial planning debates in Ireland as a means of providing a holistic social–ecological framework for spatial guidance and land use management.

4. Planning for social–ecological resilience in an Irish context

4.1. The emergence of GI

The initial thrust behind attempts to introduce the GI concept into Irish land use planning practice stemmed from a desire to remedy the perceived problem of ecosystem attrition consequent on habitat fragmentation from increasing urban-generated development in rural localities. Thus, the first formal reference to GI in an Irish policy context occurred with reference to ecological networks[3] in a study commissioned by the Irish Environmental Protection Agency (EPA) (Tubridy & O'Riain, 2002), to inform the then upcoming National Spatial Strategy (NSS) (Department of Environment, Heritage and Local Government [DOEHLG], 2002). GI was here equated with ecological networks and metaphorically explained by reference to more familiar forms of 'grey infrastructure' (transport and drainage infrastructure). With a focus on scientific principles firmly rooted in landscape ecology (Forman & Godron, 1986; Jongman & Pungetti, 2004; Wiens, 2007), GI was presented in this study as a solution to ecosystems fragmentation by creating a series of ecological 'corridors' and linking habitat 'core areas' (Tubridy & O'Riain, 2002, p. vii). In this sense, the initial interpretation and promotion of GI in Irish planning policy debates focused primarily on ecological issues with little consideration allocated to social–ecological relationships beyond the perceived detrimental influence of society on ecosystems integrity. However, the NSS when finally adopted in November of 2002 made no specific reference to the value of the ecological network ('green infrastructure') approach or its relevance to strategic planning.

Instead, the NSS advocates the development of a 'Green Structure' through regional- and county-level plans and strategies. Rather than foregrounding a concern for the conservation of biodiversity via an ecological network (i.e., GI) planning approach, the NSS 'Green Structure' approach seeks to balance polycentric urban development with a coordinated strategy for the containment of urban sprawl. Consequently, this 'Green Structure' approach shows preference for development concerns with a comparative paucity of consideration given to social–ecological interactions.

Initial progress at local authority level was piecemeal. In September 2004, South Dublin County Council (SDCC) adopted its County Development Plan for the period 2004–2010 (SDCC, 2004). The plan outlined an intention to deliver 'a Green Structure Plan for the county to identify green linkages and to allow for the intensification of use of existing and proposed amenity networks' (SDCC, 2004, p. 32), emphasising the increased use of current and proposed 'green linkages' for amenity purposes rather than habitat connectivity. A few months later in January 2005, Galway City Council (GCC) adopted its development plan for the 2005–2011 period (GCC, 2005). The recreation amenities provision policies of this plan were not included in an individual or 'community' chapter as was the normal format for such documents at the time, but rather were grouped with policies on biodiversity conservation in a chapter entitled 'Natural Heritage, Recreation and Amenity'. Tacitly suggesting that the existing integration of natural and semi-natural areas for recreational use was poor (GCC, 2005, Section 4.1), the plan sought to facilitate better integration by building on a framework presented in the previous Galway City Development Plan (1999–2005) for the establishment of a 'green network'. The 2005–2011 City Development Plan outlined how such a network offered the means by which to combine and coordinate the protection of natural heritage areas and facilitate the provision of open space for recreational purposes. One of the primary methods advocated for realising the green network was the creation of 'greenways', defined as 'pedestrian and cycle ways separated from road traffic' (GCC, 2005, Section 4.3). This presentation of the Council's green network 'greenways approach' as a means for the provision of transport, recreational and habitat connectivity echoes the language, if not necessarily the content, of both the 'green structure plan' of the South Dublin County Development Plan 2004–2010 and the ecological networks/GI approach of the 2002 EPA study. However, as opposed to the EPA study, this evolving approach increasingly sought to accommodate the social–ecological multifunctional potential of green spaces.

Adopted two months after the Galway City Development Plan, the Dublin City Development Plan 2005–2011 (DCC, 2005) echoes this shift towards a more multifunctional perspective on public open space. Indeed, Chapter 11 of the plan entitled 'Recreational Amenity and Open Space' envisaged that open space would furnish '…green chains or networks, which allow for walking and cycling and facilitate biodiversity' (DCC, 2005, p. 84). Policies contained in this plan are indicative of an inchoate change in how biodiversity conservation was conceived. This change comprised an interpretation of biodiversity as something, which like recreational amenities, can be enhanced via proactive planning, rather than simply protected by reactive designations. This change thus extends the turn towards an acknowledgement of the importance of social–ecological interactions in planning by seeking to enhance the potential positive synergies between such interactions through conscientious policy development.

4.2. The (re)emergence and evolution of 'GI'

By 2008, the desire to promote positive social–ecological interactions via multifunctional green space planning had emerged as a clearly identifiable discourse in Irish planning guidance documentation, notably in Dublin City Council (DCC, 2005; Dublin & Mid-East Regional Authorities [DRA & MERA], 2004; FCC, 2005a; GCC, 2008). The same year also witnessed the publication of the Green City Guidelines (UCD, DLRCC, FCC, & NATURA, 2008). These assert a social–ecological perspective on green space provision. In quoting Girling and Kellett (2005), these guidelines provide the first mention of GI in an Irish planning document since the EPA National study in 2002 (UCD, DLRCC, FCC, & NATURA, 2008, p. 10). However, the EPA study equated GI with the concept of an ecological network in which biodiversity protection was foregrounded on the basis of the intrinsic value of nature. In contrast, these guidelines reflect the post-2002 evolution of 'networked' concepts of land use governance by repositioning policy approaches to ecosystems from reactive protection by site designation to proactively planning for their enhancement as something of multifunctional 'value' in facilitating urban development in a manner that ensures 'our standard of living' (DOEHLG, 2008, p. 5) and 'well-being' (DCC, 2008, p. 9).

In September 2009, the Draft South Dublin Development Plan 2010–2016 (SDCC, 2009) was placed on public consultation display, and subsequently adopted in October 2010 (SDCC, 2010). Whereas the previous development plan for the area (2005–2010) promoted a 'Green Structure' that conceived a networked approach as primarily providing recreational amenities, this plan, adopted five years later, equates 'linked' and 'interconnected' open space provision as catering both for 'recreational needs' and the provision of 'valuable wildlife corridors'. Furthermore, such provision is seen as forming 'a significant green infrastructure in the County' (SDCC, 2010, p. 95). Thus, GI as a networked approach to planning is once again represented as a network of multifunctional land uses serving social and ecological requirements. Echoing the approach adopted by the Galway City Development Plan 2005–2011 (GCC, 2005), it is conceived that these 'green networks' will,

> …function as long distance walking and cycling routes as well as ecological corridors such as canals. Green networks are vital to the maintenance and facilitation of ecological corridors such as those found along major transport routes. Their main function is to link parks and other 'green' infrastructure. (SDCC, 2010, p. 96)

The suggestion here is that the function of green networks 'is to link parks' for recreational and biodiversity uses, whereas GI is perceived as something broader than these links. As such, it is implied that 'GI' subsumes recreational amenities and ecological corridors, but also includes additional land uses. Furthermore, Section 4.3 of the plan states that the Council's aim for 'Landscape, Natural Heritage and Amenities' is that this 'well defined and linked' (SDCC, 2010, p. 246) approach necessitates the development of,

> …a strategy for the creation of a Green Infrastructure for the County, promoting a balance between the protection of areas of high amenity, the facilitation of recreational use, and the provision of a network of sustainable wildlife corridors throughout the County. (SDCC, 2010, p. 246)

'Areas of high amenity' are here considered in terms of landscape aesthetics and referenced to a citation from Section 10 of the Planning and Development Act 2000–2007 (Oireachtas, 2000) regarding the onus on local authorities to '…include objectives relating to the preservation of the character of the landscape…'(SDCC, 2010, p. 246). Thus, the plan seeks to include 'the protection of areas of high amenity' with the existing pairing of recreational and ecological conservation land uses within its GI approach. This exposition indicates an evolving interpretation of the GI approach as a broadened landscape scale perspective on planning for social–ecological interactions that seeks to enhance a 'multifunctional resource' through careful planning, design and management. Additionally, the composite elements of GI are seen as nested within scalar hierarchies ranging in landform typologies and ownership attributes from 'Areas of high amenity' through to a 'network of sustainable wildlife corridors throughout the County' as well as 'allotments and private gardens'. In this sense, the GI concept increasingly served as a mechanism through which to integrate the perspectives and scalar lens of landscape research into mainstream planning practice.

4.3. The institutionalisation of GI

In April 2010, FCC issued for public consultation display its Draft County Development Plan 2011–2017 (FCC, 2010). This was subsequently adopted a year later in April 2011 (FCC, 2011). The plan includes three detailed GI maps in addition to the zoning, transport, architectural and archaeological maps normally associated with such documents. Chapter 3 of this plan is entitled 'Green Infrastructure'. The insertion of the GI chapter prior and adjacent to the subsequent conventional 'Physical Infrastructure' chapter signals an interpretation of GI as a strategically important concept binding together the various economic, physical, environmental and social objectives of the plan. The plan identifies numerous social and environmental challenges requiring redress and presents GI as a means by which to meet all these through advancing a holistic social–ecological perspective by providing,

> …space for nature (or biodiversity) and the natural systems which regulate temperature, reduce storm flows, provide us with clean water and air, and a multitude of other benefits or ecosystem services free of charge. High-quality accessible parks, open spaces and greenways provide health benefits for all … By providing a high-quality environment in which to live and to work green infrastructure helps to attract and to hold on to the high-value industries, entrepreneurs and workers needed to underpin the knowledge economy. In addition it is increasingly being recognised that green infrastructure is a vital component in building **resilient communities capable of adapting** to the consequences of climate change. [Emphasis added] (FCC, 2011, p. 91)

By specifying the 'vital' role played by GI in 'building resilient communities capable of adapting', the FCC County Development Plan advances the concept of resilience in its primary policy framework concurrent with promoting GI as the mechanism by which to facilitate such resilience. In doing so, the plan equates resilience with adaptive capacity rather than a preservation of the status quo, thereby promoting an 'evolutionary' form of social–ecological resilience.

By the summer of 2010, the GI planning policy concept appeared to be in wide circulation among a community of planning practitioners and allied professionals, with its representation evident in both regional- (DRA & MERA, 2010) and local-level planning policy guidance (DCC, 2010). GI was given further prominence by Fáilte Ireland[4] (FI, 2010), in a published document on how to maximise the tourist potential of historic towns, while reference to the GI concept in a document produced by the Heritage Council (HC, 2010) regarding the formulation of a National Landscape Strategy for Ireland, indicates a broadening perception of the approach's relevance for an array of social–ecological issues at the landscape scale. This proliferation of interpretations and references to GI continued into 2011. One of the first among these was a proposed variation to the Dún Laoghaire Rathdown County Development Plan (DLRCC, 2011) issued for public consultation in January and subsequently adopted in September of 2011. This variation presented a recreation and amenity interpretation of GI in the context of a high-density urban environment. Subsequent months saw reference made to GI within planning documentation with respect to flood risk management (Sligo County Council, 2011), long-distance walking and cycle routes, as well as with regard to ecological corridors (Athy Town Council, 2011). GI was also referenced in connection with the assessment and protection of landscape character (Department of Arts, Heritage & the Gaeltacht, 2011). Table 1 summarises and illustrates these shifting representations of GI in Irish spatial planning practice. The next section explores how employing the GI approach has helped planners and allied professionals bridge the gap between strategic policy and local practice in centralising at the local planning level the social–ecological perspective advanced by those working in the field of landscape research.

5. From strategic policy to local practice

FCC is broadly recognised as having pioneered the innovative deployment of GI planning for enhancing social–ecological resilience in Ireland (Lennon, 2013, 2014). It does so in an effort to reduce tensions between growth management and environmental protection. This entails a holistic perspective on planning that endeavours to augment the potential for social–ecological synergies that furnish quality of life enhancements while concurrently advancing ecological conservation. Such an approach also seeks to facilitate adaptation to both predictable change and unforeseen events. Thus, the GI approach advanced by FCC aims to promote an 'evolutionary' perspective on planning for the resilience of social–ecological systems.

The area administered by FCC encompasses a transition of land uses from the urban–suburban continuum extending from Dublin City to a rural coastal and agricultural landscape containing numerous European nature conservation sites designated under the provisions of the EU Birds and Habitats Directives. Realising resilience in this context is guided by a strategic approach to GI planning that advances a series of policy formulation principles. These are namely; a collaborative approach, advancing a multifunctional perspective on land use planning, as well as promoting functional and

Table 1. Evolution of GI in Irish spatial planning practice.

Timeframe	Green infrastructure as:	Key focus
Early 2000s	Ecological networks	• Ecological corridors • Linking habitats
	Green structure	• Urban growth management • Strategic greenbelts
Mid 2000s	Green linkages	• Amenity purposes
	A green network or greenways	• Protection of natural heritage areas • Provision of greenspace for recreation
	Green chains or networks	• Multifunctionality • Proactive biodiversity enhancement
Late 2000s	Multifunctional networks	• Network of multifunctional land uses serving social and ecological requirements • Landscape scale perspective • Multi-scalar
2010s	Essential infrastructure	Incorporating above • Promoting resilience and adaptation • Environmental risk management (e.g., flood risk)

spatial connectivity. The operationalisation of these principles is evidenced in innovative and interlinked local area plans for the contiguously located Baldoyle-Stapolin (FCC, 2013a) and Portmarnock South (FCC, 2013b) areas. These plans employ a GI approach to holistically frame and integrate policy initiatives concerning landscape aesthetics, biodiversity, sustainable urban drainage, archaeology and built heritage, as well as open space and recreation. Through a detailed and iterative environmental assessment process, both documents negotiate the development constraints posed by various conservation designations (SPA, SAC, Shellfish Waters) in a manner that sensitively accommodates both urban expansion and environmental protection. Included in the plans are new residential areas integrated with parkland, sustainable urban drainage schemes, non-motorised transport routes and spaces for 'urban farming' that are specifically designed to assist community development. A key feature of these plans is thus how they work synergistically in facilitating high-quality urban extensions to the Baldoyle and Portmarnock urban areas while concurrently protecting the ecological integrity of the Baldoyle Estuary. Thus, examining how FCC has developed and deployed the aforementioned series of policy formulation principles in seeking to realise social–ecological resilience in both its strategic planning objectives and the production of these local area plans furnishes insight into how the employment of a GI approach in planning practice helps centralise a landscape perspective in land use governance (Figures 1–3).

5.1. Collaborative approach

FCC is a relatively new organisation having been formed in 1994 when three new local authorities were created following the dissolution of Dublin County Council (Oireachtas, 1993). Professional staff within the council who were interviewed indicated their belief that this comparative youth stimulates an organisational identity wherein functional roles have not yet become 'sedimented' (Peters, 2005; Scott, 2008) and innovative possibilities are positively received. As noted by one interviewee, 'Fingal does innovative things. We like new thinking. We like to be able to say that about ourselves' (Interviewee A8). Such a willingness to experiment has been identified by both Ahern (2011, 2013) and Evans (2011) as essential attributes in seeking to advance social–ecological resilience. Reinforcing this identity as a dynamic local authority, FCC has undertaken a self-initiated reorganisation of its disciplinary divisions. This reorganisation was instigated with the intent of facilitating greater collaboration between the array

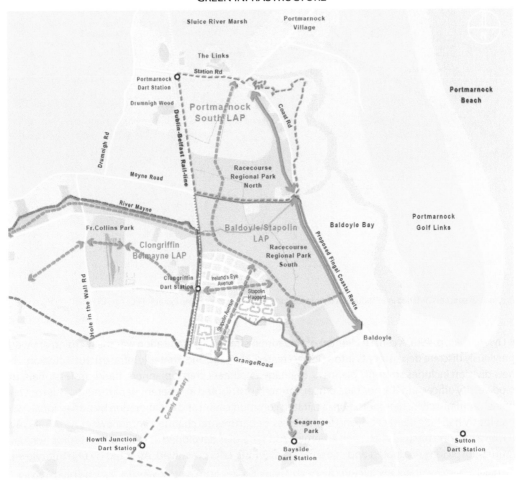

Figure 1. Baldoyle-Stapolin local area plan, GI context (source: FCC, 2013a).

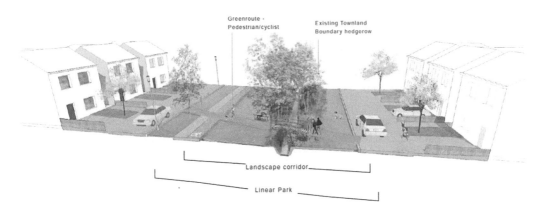

Figure 2. Linear multifunctional park concept outlined in Portmarnock South local area plan (source: FCC, 2013b).

of council professions deemed pertinent to land use planning activities. In essence therefore, it was initiated to redress the 'silo mentality' in traditional planning activities 'whereby different departments of a local authority work separately from each other—and occasionally in conflict with each other' (Kambites

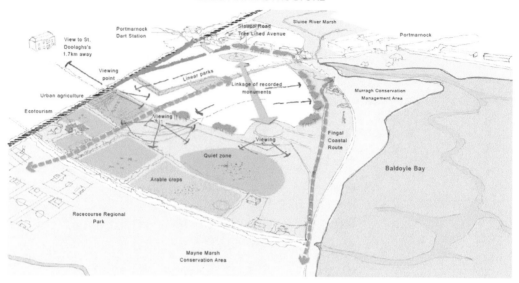

Figure 3. GI concepts outlined in introduction chapter of Portmarnock local area plan (source: FCC, 2013b).

& Owen, 2006, p. 490). A central element of this administrative reorganisation was the merging of several previously discrete departments into a newly created 'Planning and Strategic Infrastructure' division. This new division includes strategic planners, drainage engineers, traffic planners, parks professionals, the biodiversity officer and the heritage officer, formerly distributed in different departments. This root and branch administrative reorganisation facilitated communication and collaboration by professionals who previously had little contact beyond formal cross-departmental channels (Interviewees A5, A6 and A7). Positive working relationships soon emerged and synergies developed as ill-formed presumptions and mutual suspicions dissipated and cooperative planning efforts evolved. As noted by one interviewee,

> I think 'Planning and Strategic Infrastructure' makes sense. Because in the past like we would have had the Planning Department planning for things, and other Departments then delivering major infrastructure, but now you have kind of those things being thought about in a more integrated way … So the reorganisation helps I suppose in terms of making it more possible for people to come together, to talk together. So we're not as silo'ed as we were … And now I think there is much more realisation that the silos are less fixed, and so people are much more willing to talk horizontally across the organisation. (Interviewee A8)

Thus, the administrative reorganisation of FCC has advanced the potential of the local authority to plan 'in a more integrated way' by facilitating collaborative effort by a spectrum of professionals drawn from an array of theoretical backgrounds, practices and opinions (Benedict & McMahon, 2006, p. 40). Such increased 'horizontal' communication and working arrangements has helped promote more comprehensive and efficient responses to a multitude of complex planning issues by enabling concerted action in achieving seemingly disparate goals such as flood control and habitat conservation (European Commission, 2012; FCC, 2011; Novotny, Ahern, & Brown, 2010). GI has facilitated this by presenting a 'centring concept' that various professions can 'buy into' (Interviewee A8) in forging interdisciplinary collaborative working arrangements. Exemplifying FCC's openness to innovative ideas and new working relationships, it is noted that collaborative activity around the GI concept initially emerged from council officers (planners, parks professionals and the heritage officer) and not by way of instruction from senior management (Lennon, 2015).

In reflecting on the production of Baldoyle-Stapolin and Portmarnock South local area plans, those involved in overseeing policy formulation stress the role of the GI concept in focusing a diversity of practice backgrounds on potential synergies (Interviewees A1, A2, A3 and A4). In this way, GI helped stimulate collaborative engagement between professionals, and between the council and other agencies. As noted by one planner involved in the plan production process, 'Whether that is with your other Departments, or whether it was the other Agencies, it's all about collaboration' (Interviewee A4). This collaborative approach is reflected in the way the plans seek a multifunctional perspective on spatial planning, wherein each parcel of land is seen to offer the potential to serve a combination of functions, such as biodiversity conservation and flood risk management or recreation and drainage.

Moreover, the drive for innovative collaboration advanced by FCC in the development of these local area plans also involved working with local community groups through meetings and 'plebiscites' over issues of recreational need and access (Interviewee A5), as well as in monitoring the effectiveness of policy implementation. An illustrative example of such broader collaboration is the efforts by FCC to cultivate a partnership with local nature conservation NGOs to both inform policy formulation and monitor its performance. As conveyed by one interview involved in such collaborative initiatives,

> We do a lot of work with the local NGOs because they have a lot of local knowledge ... they're looking at the site for years. While a consultant comes in one or two days, makes an assessment, [and says] there's nothing there. Well they [NGOs] can say no, wait a minute; last winter there was loads of them, loads of these birds or animals and plants, whatever, they're just not here this year for whatever reason and it's more to kind of capture that and I think it requires basically a lot more interaction between nature conservation groups and the local authority. (Interviewee A2)

5.2. Multifunctionality

The significance of land use multifunctionality in the GI policy advanced by FCC is illustrated by the central 'aim' of the council's GI approach outlined in its development plan:

> Create an integrated and coherent green infrastructure for the County which will protect and enhance biodiversity, provide for accessible parks and open space, maintain and enhance landscape character including historic landscape character, protect and enhance architectural and archaeological heritage and provide for sustainable water management by requiring the retention of substantial networks of green space in urban, urban fringe and adjacent countryside areas to serve the needs of communities now and in the future including the need to adapt to climate change. (FCC, 2010, p. 89)

This strategic-level policy direction formed a departure point in the policy formulation process for the Baldoyle-Stapolin and Portmarnock South local area plans. Here, local-level policy reflects the recalibration of planning practice from traditional approaches that foster single function land uses towards a multifunctional approach that facilitates social–ecological integration. This was conveyed in the reflections of one planner involved in producing these plans when noting,

> What I think we're doing then is we're trying to provide this framework, which can be bought into by all the different parties, and which can help sustain our biodiversity, which can help make places better. It gives [us] our open spaces, our movement and all the rest. All those things that we want. ... So whereas before, while we might have been trying to do it, we didn't have this big overview, we did it a little, we wanted our park and maybe we had our habitat conservation there. And we had a cycle path over there, but we didn't put it all into that frame. So, that I suppose is maybe how I'd see it, as kind of changing the traditional. (Interviewee A1)

In comparison with conventionally produced local area plans in Ireland, these plans are atypically detailed in the provision of design guidance. It was felt that this was necessary to ensure the proper implementation of the relatively novel GI concept being advocated (Interviewee A6). Consequently, the plans detail mowing regimes, direction on how Sustainable Drainage Systems should be incorporated into the design of the public realm and guidance on public lighting so as not to cause undue interference to nocturnal animals. This multifunctional perspective on land use planning also extends into the policy construction phases of the local area plans. Here, FCC seeks to promote the use of development sites through the temporary use of undeveloped areas for social and ecological enhancement. As recounted by a council officer involved in the production of these plans,

> What we were suggesting to the developers [is] that they make all of the land accessible, except for the area that was the subject of the current phase of development, as opposed to putting up hoardings and fences. And what you do then is you cut your paths through it for cycling and walking, and then the rest of it you turn over to something like wild flower meadows or short rotation biomass, or something like that.

… and using the model like short rotation woodland or wild flower meadow, you can say to a farmer 'you've got to cut these paths seventeen times a year, and for that we'll allow you to take the hay off that area'. Or we say, 'fence off, you know with stock proof fencing, Phase B, and the Council will graze it with an attractive set of rare breeds, or something like that'. So you can create something that is attractive, sustainable, and easy to manage, as an interim to the final development of the site. (Interviewee A6)

5.3. *Connectivity*

The collaborative approach that facilitates multifunctional synergies has also facilitated more attention to spatial and functional connectivity between land uses in local policy formulation and implementation. Prior to the advocacy of a GI planning approach, FCC had advanced habitat connectivity via ecological networks (FCC, 2005b). Such networks render otherwise fragmented ecosystems biologically coherent by facilitating species movement and genetic exchange (Opdam, Steingröver, & Rooij, 2006; Pungetti & Romano, 2004). Although promoting spatial and scalar integration, these networks focused primarily on 'ecological' connectivity. Consequently, this wholly ecological focus failed to fully reflect the social dynamics intrinsic to social–ecological systems thinking. However, following greater acquaintance with GI theory and the consequent advocacy of a holistic approach to planning, FCC has sought to advance a more functionally integrated network of key sites that meet several social objectives while concurrently maintaining ecosystems integrity. This GI network is given graphic representation in a series of planning maps accompanying the County Development Plan that identify key sites of conservation and amenity value linked via a series of multipurpose corridors. A key aspect of planning this GI network has been the use of spatial data analysis in identifying opportunities for enhanced connectivity. Using such evidence, efforts are made to produce comprehensive maps of GI assets from which to formulate site-specific initiatives that consolidate the broader GI network. However, Kambites and Owen (2006, p. 488) advise that if such cartographic exercises are 'not set within an effective planning process, the mapping of green infrastructure, albeit a vital component of the process, remains little more than a technical exercise'. Accordingly, FCC officers express an understanding that mapping GI assets is a means to an end rather than an end in itself. In this sense, the maps employed to assist planning policy formulation form tools which aid rather than replace critical engagement with a GI planning approach. Engaging with this approach ultimately requires promoting synergistic social–ecological integration by focusing on how the multifunctional potential of GI networks can be sensitively realised. As noted by one interviewee when reflecting on FCC's GI planning approach,

It's [GI] basically trying to link up your key ecological features which are amenity features, your water features and the likes of that…

…most of the important major conservation in the county is within this network so if you're going to do any development near it, whether it's amenity or whether it is roads or water or housing, these are the key features that need to be protected and it's more to see how can we work with you to incorporate that. If you build a housing estate and the river runs through that, how can we design the flood plain at the river in such a way that it will actually suit everybody. So it is still an amenity space, but wildlife can live there too … it's trying to combine those different things. (Interviewee A2)

This approach is reflected within the Baldoyle-Stapolin and Portmarnock South local area plans. Here connectivity is promoted both within the plan lands and with contiguous land uses. Such a perspective is given prominence in the 'Overarching Green Infrastructure Strategy' for the Baldoyle-Stapolin Local Area Plan which states,

This LAP seeks to create a green infrastructure network of high quality amenity and other green spaces that permeate through the plan lands while incorporating and protecting the natural heritage and biodiversity value of the lands. (FCC, 2013a, p. 18)

Illustrated in this strategic objective is a desire to integrate both the biological focus of ecological networks with the social concerns of greenways to deliver multifunctional connectivity (Austin, 2014; Benedict & McMahon, 2006; Rouse & Bunster-Ossa, 2013). In this sense, FCC has sought to employ a broad-based collaborative approach to facilitate multifunctionality and connectivity across the urban–rural interface in a sensitive ecological context wherein there exists significant pressure for urban expansion. The council has endeavoured to do so by deploying a GI planning approach to centralise the holistic perspective of landscape research that promotes social–ecological resilience in acknowledging the mutually constitutive nature of ecosystems health and human well-being.

6. Conclusion

GI has increasingly become an established policy discourse at regional and local levels of the planning hierarchy in Ireland since 2008. The emergence, ongoing evolution and widening institutionalisation of the GI approach indicate a growing centralisation of landscape perspectives in Irish planning practice. However, GI-specific planning guidance at a national level is conspicuous by its absence. Consequently, the GI planning approach in Ireland is primarily employed at the local authority level with a more strategically GI informed landscape approach evident in some, but not all, regional guidance. In this sense, county- and city-level development plans have emerged as the primary vehicle through which GI guidance is formulated and a holistic social–ecological (landscape) perspective on resilience planning is integrated into land use policy. The strategic direction provided by such policy is then given site base application in local area plans wherein the details on how to deliver social–ecological resilience are developed. Nevertheless, there are variations in the interpretation and application of the GI concept between local authorities. Several local authority plans demonstrate a prioritisation of GI for biodiversity protection, but seek to partially advance a more multifunctional approach to conservation by including recreational open space provision within policies concerning natural heritage management (Kildare County Council, 2012). However, many of those local authorities employing the GI concept exercise it as an extension rather than a transformation of traditional approaches to environmental conservation (Meath County Council, 2013; Monaghan County Council, 2013). In such instances, GI may be conceived as a re-branding of single use 'ecological-networks' akin to that advanced in the study commissioned by the EPA in 2002 (Tubridy & O'Riain, 2002). Envisaging GI in such a manner confines it to biodiversity conservation. Consequently, these interpretations risk eroding the holistic social–ecological perspective of GI that seeks to advance the synergistic multifunctional potential of land uses. Here, issues like flood management, accessible green space provision and non-motorised transport may be perceived in a disjointed fashion as a restricted GI approach is formulated to accord with existing administrative delineations. This phenomenon can be witnessed in the sustained configuration of development plans wherein 'natural heritage' is confined to a distinct plan chapter that is frequently disengaged from other issue-specific policies, such as 'drainage' and 'transport'. In the absence of a section at the beginning of a plan to first outline how a GI approach structures subsequent chapters and polices (FCC, 2011), maintaining the conventional structure of plans in this fashion reinforces existing administrative compartmentalisation and reduces the transformative potential of the GI concept to facilitate the synergistic integration of land uses and the promotion of social–ecological resilience. To date, this phenomenon seems most pronounced in Irish rural local authorities whose capacity to fully engage a proactive multifunctional GI planning approach may be hampered by resource constraints such as a skills deficit, low staffing and restricted budgets.

In contrast, FCC has been to the fore in Ireland in seeking to advance the GI planning approach. At the heart of the FCC's activities is a drive to enhance collaborative working arrangements to encourage a more responsive and effective holistic approach to the complexities of planning for social–ecological resilience. This paper's review of FCC's efforts to promote such a perspective illustrates how the theory of GI has been used as a 'centring concept' (Interviewee A8) that stimulates inter-disciplinary working to enable the formulation of an 'organisational strategy that provides a framework for planning conservation and development' (Benedict & McMahon, 2006, p. 15). With a focus on improving the multifunctional potential of connected local and landscape scale environmental assets (Davies, Macfarlane, & Roe, 2006; Lafortezza, Davies, Sanesi, & Konijnendijk, 2013), such a GI approach supplies 'the "umbrella" for disciplines to unite' (Wright, 2011, p. 1011) and consequently promotes 'increased dialogue between planners, developers, and policy-makers' (Mell, 2010, p. 241).

However, we caution this with an awareness that the history of planning is littered with the carcasses of failed 'blueprints' (Ostrom, Janssen, & Anderies, 2007) that proposed a universally applicable solution to delivering on the promise of sustainability (Baker & Eckerberg, 2008; Owens & Cowell, 2011). Indeed, continuing dispute on how planning should seek to advance more sustainable forms of governance indicates ongoing failure in the search for a single means to resolve persistent divergence between environmental protection, economic development and social equity (Allmendinger, 2009; Carter, 2007; Torgerson & Paehlke, 2005). This issue is intensified in an Irish context wherein there is an 'implementation deficit' as the planning practice of GI policy formulation largely awaits the planning practicalities of translation into evaluable material change. Thus, we do not claim that GI furnishes a panacea for the multitude of problematic issues encountered in planning practice. Rather, what this paper demonstrates is that progressing a landscape perspective in planning necessitates an openness to new ideas and new ways of working wherein cognizance of knowledge limitations promotes 'learning to manage by managing to learn' (Bormann, Cunningham, Brookes, Manning, & Collopy, 1994, p. 1). Key to this is overcoming the 'silo approach to planning' through 'a transformation of the structural context and factors that determine the frame of reference' for planning activity (Pahl-Wostl, 2009, p. 359). Accordingly, integrating a more landscape informed holistic perspective on social–ecological resilience requires the 'recognition that multiple sources and types of knowledge are relevant to problem solving' (Armitage et al., 2008, p. 96). This foregrounding of inclusivity resonates with other moves in planning theory that seek to ground planning in a more 'collaborative' ethos (Agger & Lofgren, 2008; Healey, 2003; Innes & Booher, 2010) as a means to resolve conflict through cooperation and the accommodation of difference (Forester, 1999; Umemoto & Igarashi, 2009). In this sense, a planning perspective better attuned to landscape research requires collaborative learning (Goldstein, 2009), and experimentation (Ahern, 2011), wherein social–ecological 'systems' are seen to be co-produced and co-evolve with forms of locally grounded scientific-administrative knowledge (Evans, 2011).

Notes

1. Section 19 of the Planning and Development (Amendment) Act 2012 specifies that this review interval can be deferred subject to provisions specified in the Act regarding deferral time limits and the justifications for seeking a deferral.
2. Working on a broad definition of 'heritage', these officers help coordinate and provide input to numerous council activities ranging from natural environmental issues through to landscape and archaeology, as well as built and cultural heritage matters. As such, their activities frequently interact with the local planning policy development process.
3. Defined by Tubridy and O'Riain, (Tubridy & O'Riain, 2002, p. 1) as, 'a network of sites. Its constituents are: "core areas" of high biodiversity value and "corridors" or "stepping stones", which are linkages between them. In contrast to species or site based conservation, the ecological network approach promotes management of "linkages" between areas of high biodiversity value, between areas of high and low biodiversity value, between areas used by species for different functions, and between local populations of species. "Corridors" or linking areas can support species migration, dispersal or daily movements'.
4. Ireland's National Tourism Development Authority.

Disclosure statement

No potential conflict of interest was reported by the authors.

References

Agger, A., & Lofgren, K. (2008). Democratic assessment of collaborative planning processes. *Planning Theory, 7*, 145–164.
Ahern, J. (2011). From fail-safe to safe-to-fail: Sustainability and resilience in the new urban world. *Landscape and Urban Planning, 100*, 341–343.
Ahern, J. (2013). Urban landscape sustainability and resilience: The promise and challenges of integrating ecology with urban planning and design. *Landscape Ecology, 28*, 1203–1212.

Ahern, J., Cilliers, S., & Niemelä, J. (2014). The concept of ecosystem services in adaptive urban planning and design: A framework for supporting innovation. *Landscape and Urban Planning, 125,* 254–259.

Allmendinger, P. (2009). *Planning theory.* Hampshire: Palgrave Macmillan.

Armitage, D. R., Plummer, R., Berkes, F., Arthur, R. I., Charles, A. T., Davidson-Hunt, I. J., … Wollenberg, E. K. (2008). Adaptive co-management for social–ecological complexity. *Frontiers in Ecology and the Environment, 7,* 95–102.

Athy Town Council. (2011). *Draft Athy town plan 2012–2018.* Athy, Co. Kildare: Author.

Austin, G. (2014). *Green infrastructure for landscape planning: Integrating human and natural systems.* London: Routledge.

Baker, S., & Eckerberg, K. (Eds.). (2008). *In pursuit of sustainable development: New governance practices at the sub-national level in Europe.* Oxford: Taylor & Francis Group.

Barr, S., & Devine-Wright, P. (2012). Resilient communities: Sustainabilities in transition. *Local Environment, 17,* 525–532.

Benedict, M., & McMahon, E. (2006). *Green infrastructure: Linking landscapes and communities.* London: Island Press.

Benson, J. F., & Roe, M. (Eds.). (2007). *Landscape and sustainability.* New York, NY: Spon Press.

Berg, B. L. (2004). *Qualitative research methods.* London: Pearson Education.

Bormann, B. T., Cunningham, P. G., Brookes, M. H., Manning, V. W., & Collopy, M. W. (1994). *Adaptive ecosystem management in the Pacific Northwest* (Gen. Tech. Rep. PNW-GTR-341). Portland, OR: US Department of Agriculture, Forest Service: Pacific Northwest Research Station.

Bryman, A. (2008). *Social research methods.* Oxford: Oxford University Press.

Carter, N. (2007). *The politics of the environment.* Cambridge: Cambridge University Press.

Coates, P. (1998). *Nature: Western attitudes since ancient times.* Cambridge: Polity Press.

Collier, M. J., Nedović-Budić, Z., Aerts, J., Connop, S., Foley, D., Foley, K., … Verburg, P. (2013). Transitioning to resilience and sustainability in urban communities. *Cities, 32*(Suppl. 1), S21–S28.

Cumming, G. S. (2011). *Spatial resilience in social-ecological systems.* London: Springer.

Davidson, D. J. (2010). The applicability of the concept of resilience to social systems: Some sources of optimism and nagging doubts. *Society & Natural Resources, 23,* 1135–1149.

Davies, C., Macfarlane, R., & Roe, M. H. (2006). *Green infrastructure planning guide, 2 volumes: Final report and GI planning.* Newcastle: University of Northumbria, North East Community Forests, University of Newcastle, Countryside Agency, English Nature, Forestry Commission, Groundwork Trusts.

Davoudi, S., Shaw, K., Haider, L. J., Quinlan, A. E., Peterson, G. D., Wilkinson, C., … Porter, L. (2012). Resilience: A bridging concept or a dead end? "Reframing" resilience: Challenges for planning theory and practice Interacting Traps: Resilience assessment of a pasture management system in northern Afghanistan Urban resilience: What does it mean in planning practice? Resilience as a useful concept for climate change adaptation? The politics of resilience for planning: A cautionary note. *Planning Theory & Practice, 13,* 299–333.

Dawley, S., Pike, A., & Tomaney, J. (2010). Towards the resilient region? *Local Economy, 25,* 650–667.

DCC. (2005). *Dublin city development plan 2005–2011.* Dublin: Author.

DCC. (2008). *Dublin city biodiversity action plan 2008–2012.* Dublin: Author.

DCC. (2010). *Dublin city development plan 2011–2017.* Dublin: Author.

Department of Arts, Heritage and the Gaeltacht. (2011). *Strategy issues paper for consultation for a National Landscape Strategy for Ireland.* Dublin: Author.

Department of Environment, Community and Local Government. (2015). *Planning policy statement.* Dublin: Author.

DLRCC. (2011). *Proposed variation no. 2 to the Dún Laoghaire Rathdown County development plan 2010–2016 (Sandyford Urban Framework Plan).* Dublin: Author.

DOEHLG (Department of Environment, Heritage and Local Government). (2002). *National spatial strategy for Ireland, 2002–2020.* Dublin: Author.

DOEHLG (Department of Environment, Heritage and Local Government). (2008). *The economic and social aspects of biodiversity: Benefits and costs of biodiversity in Ireland.* Dublin: Author.

DRA & MERA. (2004). *Regional planning guidelines for the greater Dublin area.* Dublin: Author.

DRA & MERA. (2010). *Regional planning guidelines for the greater Dublin area 2010–2022.* Dublin: Author.

Eraydin, A., & Taşan-Kok, T. (Eds.). (2012). *Resilience thinking in urban planning.* London: Springer.

Erixon, H., Borgström, S., & Andersson, E. (2013). Challenging dichotomies—Exploring resilience as an integrative and operative conceptual framework for large-scale urban green structures. *Planning Theory & Practice, 14,* 349–372.

European Commission. (2012). *The multifunctionality of green infrastructure.* Brussels: Author.

Evans, J. P. (2011). Resilience, ecology and adaptation in the experimental city. *Transactions of the Institute of British Geographers, 36,* 223–237.

FCC. (2005a). *Fingal County development plan 2005–2011.* Dublin: Author.

FCC. (2005b). *Fingal County heritage plan 2005–2010.* Dublin: Author.

FCC. (2010). *Draft Fingal County development plan 2011–2017.* Dublin: Author.

FCC. (2011). *Fingal County development plan 2011–2017.* Dublin: Author.

FCC. (2013a). *Baldoyle-Stapolin local area plan.* Swords, Dublin: Author.

FCC. (2013b). *Portmarnock South local area plan.* Swords, Dublin: Author.

FI. (2010). *Historic towns in Ireland: Maximising your tourist potential.* Dublin: Author.

Folke, C. (2006). Resilience: The emergence of a perspective for social–ecological systems analyses. *Global Environmental Change, 16,* 253–267.

Folke, C., Carpenter, S. R., Walker, B., Scheffer, M., Chapin, T., & Rockström, J. (2010). Resilience thinking: Integrating resilience, adaptability and transformability. *Ecology and Society, 15*, 20. Retrieved from http://www.ecologyandsociety.org/vol15/iss4/art20

Folke, C., Colding, J., & Berkes, F. (2003). Synthesis: Building resilience and adaptive capacity in social-ecological systems. In F. Berkes, J. Colding, & C. Folke (Eds.), *Navigating social-ecological systems: Building resilience for complexity and change* (pp. 352–387). Cambridge: Cambridge University Press.

Forester, J. (1999). *The deliberative practitioner: Encouraging participatory planning processes.* Cambridge, MA: MIT Press.

Forester, J. (2013). On the theory and practice of critical pragmatism: Deliberative practice and creative negotiations. *Planning Theory, 12*, 5–22.

Forman, R. T. T., & Godron, M. (1986). *Landscape ecology.* New York, NY: Wiley.

Foucault, M. (1972). *The archaeology of knowledge.* London: Routledge.

Fünfgeld, H., & McEvoy, D. (2012). Resilience as a useful concept for climate change adaptation. *Planning Theory & Practice, 13*, 324–328.

GCC. (2005). *Galway city development plan 2005–2011.* Galway: Author.

GCC. (2008). *Galway city recreational and amenity needs study.* Galway: Author.

Geertz, C. (1973). *The interpretation of cultures.* New York, NY: Basic Books.

Girling, S., & Kellett, R. (2005). *Skinny streets and green neighbourhoods: Design for environment and community.* Washington, DC: Island Press.

Glaser, M., Krause, G., Ratter, B. M. W., & Welp, M. (2012). *Human–nature interactions in the anthropocene: Potentials of social-ecological systems analysis.* New York, NY: Taylor & Francis.

Goldstein, B. (2009). Resilience to surprises through communicative planning. *Ecology and Society, 14*, 33. Retrieved from http://www.ecologyandsociety.org/vol14/iss2/art33/

Goudie, A. S. (2009). *The human impact on the natural environment: Past, present, and future.* Oxford: Wiley.

Gunderson, L. H., & Holling, C. S. (2001). *Panarchy: Understanding transformations in human and natural systems.* Washington, DC: Island Press.

Haider, L. J., Quinlan, A. E., & Peterson, G. D. (2012). Interacting traps: Resilience assessment of a pasture management system in northern Afghanistan. *Planning Theory & Practice, 13*, 312–318.

HC. (2010). *Proposals for Ireland's landscapes.* Kilkenny: Author.

Healey, P. (2003). Collaborative planning in perspective. *Planning Theory, 2*, 101–123.

Holling, C. S. (1973). Resilience and stability of ecological systems. *Annual Review of Ecology and Systematics, 4*, 1–23.

Ingold, T. (2000). *The perception of the environment.* New York, NY: Routledge.

Innes, J. E., & Booher, D. E. (2010). *Planning with complexity: An introduction to collaborative rationality for public policy.* New York, NY: Taylor & Francis.

Jongman, R. H. G., & Pungetti, G. (Eds.). (2004). *Ecological networks and greenways; conception, design, implementation.* Cambridge: Cambridge University Press.

Kambites, C., & Owen, S. (2006). Renewed prospects for green infrastructure in the UK. *Planning Practice and Research, 21*, 483–496.

Kildare County Council. (2012). *Kildare town local area plan.* Naas: Author.

Kvale, S. (1996). *InterViews: An introduction to qualitative research interviewing.* London: Sage.

Lafortezza, R., Davies, C., Sanesi, G., & Konijnendijk, C. (2013). Green infrastructure as a tool to support spatial planning in European urban regions. *iForest - Biogeosciences and Forestry, 6*, 102–108.

Lennon, M. (2013). *Meaning making and the policy process: The case of green infrastructure planning in the Republic of Ireland* (Unpublished doctoral dissertation). Cardiff University, UK.

Lennon, M. (2014). Green infrastructure and planning policy: A critical assessment. *Local Environment, 20*, 957–980. doi:10.1080/13549839.2014.880411

Lennon, M. (2015). Explaining the currency of novel policy concepts: Learning from green infrastructure planning. *Environment and Planning C: Government and Policy, 33*, 1039–1057.

Lennon, M., & Scott, M. (2014). Delivering ecosystems services via spatial planning: Reviewing the possibilities and implications of a green infrastructure approach. *Town Planning Review, 85*, 563–587.

Lentzos, F., & Rose, N. (2009). Governing insecurity: Contingency planning, protection, resilience. *Economy and Society, 38*, 230–254.

McHarg, I. L. (1969). *Design with nature.* Oxford: Wiley.

Meath County Council. (2013). *Meath county development plan 2013–2016.* Navan, Co. Meath: Author.

Mell, I. C. (2010). *Green infrastructure: Concepts, perceptions and its use in spatial planning* (Unpublished doctoral dissertation). Newcastle University, UK.

Monaghan County Council. (2013). *Monaghan county development plan 2013–2019.* Monaghan: Author.

Norris, F., Stevens, S., Pfefferbaum, B., Wyche, K., & Pfefferbaum, R. (2008). Community resilience as a metaphor, theory, set of capacities, and strategy for disaster readiness. *American Journal of Community Psychology, 41*, 127–150.

Novotny, V., Ahern, J., & Brown, P. (2010). *Water centric sustainable communities: Planning, retrofitting and building the next urban environment.* Hoboken, NJ: Wiley.

Oireachtas. (1993). Local Government (Dublin) Act. *Number 31 of 1993.* Ireland: Author.

Oireachtas. (2000). The Planning and Development Act. *No. 30 of 2000*. Ireland: Government Publications Office.

Opdam, P., Steingröver, E., & Rooij, S. V. (2006). Ecological networks: A spatial concept for multi-actor planning of sustainable landscapes. *Landscape and Urban Planning, 75*, 322–332.

Ostrom, E., Janssen, M. A., & Anderies, J. M. (2007). Going beyond panaceas. *Proceedings of the National Academy of Sciences, 104*, 15176–15178.

Owens, S., & Cowell, R. (2011). *Land and limits: Interpreting sustainability in the planning process*. New York, NY: Routledge.

Pahl-Wostl, C. (2009). A conceptual framework for analysing adaptive capacity and multi-level learning processes in resource governance regimes. *Global Environmental Change, 19*, 354–365.

Pendall, R., Foster, K. A., & Cowell, M. (2010). Resilience and regions: Building understanding of the metaphor. *Cambridge Journal of Regions, Economy and Society, 3*, 71–84.

Peters, G. B. (2005). *Institutional theory in political science: The "New Institutionalism"*. London: Continuum.

Pickett, S. T. A., Cadenasso, M. L., & Grove, J. M. (2004). Resilient cities: Meaning, models, and metaphor for integrating the ecological, socio-economic, and planning realms. *Landscape and Urban Planning, 69*, 369–384.

Pike, A., Dawley, S., & Tomaney, J. (2010). Resilience, adaptation and adaptability. *Cambridge Journal of Regions, Economy and Society, 3*, 59–70.

Plieninger, T., & Bieling, C. (2012a). Connecting cultural landscapes to resilience. In T. Plieninger & C. Bieling (Eds.), *Resilience and the cultural landscape* (pp. 3–26). Cambridge: Cambridge University Press.

Plieninger, T., & Bieling, C. (2012b). *Resilience and the cultural landscape*. Cambridge: Cambridge University Press.

Pungetti, G., & Romano, B. (2004). Planning the future landscape between nature and culture. In R. H. G. Jongman & G. Pungetti (Eds.), *Ecological networks and greenways; conception, design, implementation* (pp. 107–127). Cambridge: Cambridge University Press.

Rouse, D. C., & Bunster-Ossa, I. F. (2013). *Green infrastructure: A landscape approach*. Washington, DC: American Planning Association.

Rydin, Y. (2007). Re-examining the role of knowledge within planning theory. *Planning Theory, 6*, 52–68.

Sligo County Council. (2011). *Sligo county development plan 2011–2017*. Sligo: Author.

Scott, W. R. (2008). *Institutes and organisations: Ideas and interests*. London: Sage.

Scott, M. (2013). Resilience: A conceptual lens for rural studies? *Geography Compass, 7*, 597–610.

SDCC. (2004). *South Dublin county development plan 2004–2010*. Dublin: Author.

SDCC. (2009). *Draft South Dublin county development plan 2010–2016*. Dublin: Author.

SDCC. (2010). *South Dublin county development plan 2010–2016*. Dublin: Author.

Selman, P. (2012). *Sustainable landscape planning: The reconnection agenda*. Abingdon: Routledge.

Shaw, K. (2012). The rise of the resilient local authority? *Local Government Studies, 38*, 281–300.

Shaw, K., & Maythorne, L. (2013). Managing for local resilience: Towards a strategic approach. *Public Policy and Administration, 28*, 43–65.

Spirn, A. W. (1984). *The granite garden: Urban nature and human design*. London: Basic Books (Penguin).

Steiner, F. R. (2002). *Human ecology: Following nature's lead*. Washington, DC: Island Press.

Steiner, F. R. (2008). *The living landscape: An ecological approach to landscape planning*. Washington, DC: Island Press.

Teigão dos Santos, F., & Partidário, M. R. (2011). SPARK: Strategic planning approach for resilience keeping. *European Planning Studies, 19*, 1517–1536.

Torgerson, D., & Paehlke, R. (Eds.). (2005). *Managing Leviathan: Environmental politics and the administrative state*. Plymouth: Broadwiew Press.

Tubridy, M., & O'Riain, G. (2002). *Preliminary study of the needs associated with a National Ecological Network*. Wexford: Environmental Protection Agency.

UCD, DLRCC, FCC, & NATURA. (2008). *Green city guidelines*. Dublin: UCD Urban Institute.

Umemoto, K., & Igarashi, H. (2009). Deliberative planning in a multicultural milieu. *Journal of Planning Education and Research, 29*, 39–53.

Wagenaar, H., & Wilkinson, C. (2013). Enacting resilience: A performative account of governing for urban resilience. *Urban Studies, 52*, 1265–1284.

Walker, B. H., Gunderson, L. H., Kinzig, A. P., Folke, C., Carpenter, S. R., & Schultz, L. (2006). A handful of heuristics and some propositions for understanding resilience in social-ecological systems. *Ecology and Society, 11*, 13. Retrieved from http://www.ecologyandsociety.org/vol11/iss1/art13/

Walker, B., & Salt, D. (2006). *Resilience thinking: Sustaining ecosystems and people in a changing world*. Washington, DC: Island Press.

Wiens, J. A. (2007). *Foundation papers in landscape ecology*. New York, NY: Columbia University Press.

Wilkinson, C. (2012a). Social-ecological resilience: Insights and issues for planning theory. *Planning Theory, 11*, 148–169.

Wilkinson, C. (2012b). Urban resilience: What does it mean in planning practice? *Planning Theory and Practice, 13*, 319–324.

Wright, H. (2011). Understanding green infrastructure: The development of a contested concept in England. *Local Environment, 16*, 1003–1019.

Wylie, J. (2005). A single day's walking: Narrating self and landscape on the South West Coast Path. *Transactions of the Institute of British Geographers, 30*, 234–247.

Urban green infrastructure and urban forests: a case study of the Metropolitan Area of Milan

Giovanni Sanesi ⓘ, Giuseppe Colangelo ⓘ, Raffaele Lafortezza, Enrico Calvo and Clive Davies

ABSTRACT

Rapid expansion of urban built-up areas since the 1950s has led to the Milan region becoming one of the major metropolitan areas of Europe. This has been accompanied by significant structural changes to urban and peri-urban landscapes and fragmentation of formerly contiguous green corridors by the distribution of new urban forms such as housing and transport infrastructure. The need to address the loss of green space was first recognised by policy-makers at the end of the 1970s and in due course, this has led to new policies and laws. These policies included the introduction of the Milan metropolitan parks approach that, nowadays, is represented by numerous urban forests that have become the backbone of green infrastructure (GI) creation and management. In the last decades, a total of 10 000 hectares of new forests and green systems have been created. Boscoincittà and Parco Nord Milano are the best known examples of this approach aimed to redevelop the neighbourhoods of some suburbs of Milan to create multifunctional green spaces (forests, grasslands, wetlands, river corridor, and allotment gardens) in lands previously industrial or uncultivated. The creation and management of urban forests has become the backbone of GI creation and management in the Metropolitan Area of Milan. In recent decades, trends of land use change have been characterised by a rapid decrease in natural and agricultural areas and an increase in artificial and urban structures. Although the phenomenon is growing rapidly in this area, there is evidence of an opposite social and environmental trend highlighting the importance of GI positively affecting urban quality of life. Recent policies and management plans are dealing with this evidence by turning their attention to expanding green areas and infrastructure. The purpose of our investigation is to revisit effective measures designed to increase the quality and quantity of UGI in the metropolitan region under study. To this end, we assessed land use changes and described the potentialities and impacts of policies on such phenomena. The study analyses the main elements of UGI in the Italian context within the framework of the European Union Life + project called Emonfur, a research programme involving, *inter alia*, the establishment of an Urban Forest inventory and impact analysis of ecosystem services in the Metropolitan Area of Milan. Our research has allowed us to determine the current status of key sites by monitoring the policy and planning decisions that resulted in their development. We believe that such an analysis can pave the way to understand future land-use dynamics not only in northern Italy but in other metropolitan territories as well.

Introduction

Urban land-use developments of all sorts have great impacts on the landscape. Such developments change the content, form and function of urban landscapes and, as a consequence, the ecosystem services they provide. In particular, urban development normally reduces the area of agriculture and forestry, which in turn decreases ecosystem values due to progressive soil sealing. Urban development also fragments the regional territory and leads to a loss of connectivity between habitats. This loss is made especially evident by unsympathetic construction of urban infrastructure, although in some cases, a positive by-product is the unintended creation of new green corridors such as the verge areas of motorways and railways. Degraded ecosystems are characterised by a reduced variety of habitats, fewer species and poorer regulatory ecosystem services (Collingham & Huntley, 2000; Geneletti, 2004).

Strategic level planning at the city and regional scales can limit and counteract the loss of ecosystem services through the creation of a network of natural and semi-natural areas. This planning approach is presently referred to as green infrastructure (GI) planning (Benedict & McMahon, 2002; Davies et al., 2015; Forman, 1995; Lafortezza, Davies, Sanesi, & Konijnendijk, 2013; Mell, 2010; Weber et al., 2006). When planning is mainly concerned with urban areas, as is the case in the Metropolitan Area of Milan, the terminology urban green infrastructure (UGI) is generally applied (Hansen & Pauleit, 2014).

Cities and their regions can be considered as urban ecosystems with a high population density. Over 54% of the world's population lives in urban areas; the share in Europe is 73% and 69% in Italy (UN, 2014). In cities and their surrounding regions, the conflict between 'artificiality' and 'naturalness' is at maximum.

Typical UGI components include urban parks, protected natural areas, agricultural lands, urban-influenced forests and recreational green spaces. This has currently become the focus of a major research project, GREEN SURGE (Green Infrastructure and Urban Biodiversity for Sustainable Urban Development and the Green Economy). The project focuses on multifunctionality as the central theme for arguing the economic (and other) value of green areas in the urban context. It is based on a collaboration among 24 partners in 11 countries, funded through the European Union's Seventh Framework Programme. Project partners have identified 40 components of UGI. However, only when the components are considered as part of a system, and especially when they are inter-connected, they can be regarded as GI. Even in the absence of strategic planning, UGI delivers by default many services ranging from the preservation of cultural heritage sites to climate change adaptation to the conservation of natural areas and species (e.g., Benedict & McMahon, 2002; Lafortezza et al., 2013). However, the delivery of such services in the absence of a strategic framework is less than optimal as UGI can be designed to perform specific functions and solve local problems such as improving the hydrological response of soils or the production of food (urban agriculture). As opposed to GI, UGI is still a new area of research and much remains to be understood. It is for this reason that it is the topic of investigation.

There are many GI projects, some of which are urban focused, already at an advanced stage worldwide, especially in the USA (Lovell & Taylor, 2013; Young, 2011; Young & McPherson, 2013). In Europe, there is, for example, the Community Forests in England, which is one of a number of UK-based policy instruments that deliver ecosystem functions across municipal boundaries and re-instate damaged landscapes (Mell, 2009; Roe & Mell, 2013). Spain has also made similar efforts in this direction, for example, the 'Anela verda' in Barcelona includes a network of 12 protected areas located around the city, which are connected by ecological corridors (Lafortezza et al., 2013; Llausàs & Roe, 2012). The Ruhrgebiet, one of Europe's most populous metropolitan regions and the largest in Germany with over 11 million inhabitants in North Rhine Westphalia, is justifiably famous for its 'industrial nature' sites, which have become a tourist attraction and are linked to preserving its cultural heritage; this approach evolved at the end of the last century (Blotevogel, 1998). Portugal also presents some interesting cases of GI (Madureira & Andresen, 2014; Madureira, Andresen, & Monteiro, 2011).

In Italy, there are comparatively few examples of the delivery of GI either in rural or urban contexts. Within the urban domain, the well-established Green Belt of Turin (Cassatella, 2013), the GI approach in Milan and the Green Belt of Mirandola (Modena) are among few examples; but the Modena example is primarily aimed at climate mitigation and reducing energy consumption rather than multifunctionality.

Other examples include planned or current interventions on the Arno River in Tuscany and the potential for UGI in Rome is discussed by Barbati, Corona, Salvati, and Gasparella (2013).

Because of the availability of data, land-use trends and the long duration of research, the Metropolitan Area of Milan provides a case study of the processes of urban development, loss of land, fragmentation and reduction in the resilience of ecosystems. This area could be considered a leader in activating policies and management processes focusing on promoting UGI in contrast with other Italian regions. This study reviews the efficacy of measures designed to increase the quality and quantity of UGI given that oftentimes, Italian regions are characterised by different social and economic conditions. From a technical point of view, we deemed it important to analyse the metropolitan region and draw findings that would interest researchers and practitioners alike, especially those interested in growing metropolitan areas. The goals of the present study are threefold: (i) to highlight the changes in land use that have characterised the Metropolitan Area of Milan from 1954 to the present; (ii) to identify the legislative framework and social and economic dynamics that have influenced such land-use changes; (iii) to evaluate the efficacy of the UGI policy and planning.

The Milan Metropolitan Area

The Metropolitan Area of Milan is located within the Lombardy Region of Northern Italy. The economic output of the region is one of the highest in Europe, expressed in Gross Domestic Product (GDP) per capita (EU, 2014). There has been a continual process of urban development (European Environment Agency [EEA], 2006) causing fragmentation of the landscape (EEA, 2006). Significant changes in land use are also occurring as the economy moves from being industrial to service-based, a process which is now well advanced.

The first Italian National Report on 'Loss of Land' (ONCS, 2009) reveals that approximately 14% of the regional surface is now covered by urban areas (288,000 hectares: 310 m^2 per inhabitant) and that in the period from 1999 to 2006, the loss of agricultural land was about 27 000 hectares. According to ISPRA data (ISPRA, 2014), this loss further increased in 2009 (8.7–12.1%) and 2012 (8.8–12.4%) in terms of the estimate of soil consumed in proportion to the regional surface per year. In addition, there is a lack of available green space in the Metropolitan Area of Milan as it is surrounded by more densely built-up areas when compared to other North European cities (Kasanko et al., 2006). In a broad sense, Lombardy is a good example of a European region that is highly disturbed by human activity; for example, forests have been altered in their structure and distribution and reduced to small patches surrounded by agricultural and urban areas. Table 1 illustrates the metrics and indices applied to the spatial pattern analysis (FRAGSTATS) aimed at quantifying the forestry resources in the study area. Landscape metrics describe the existing relationship and patterns complexity of forestry patches in terms of composition and spatial distribution (McGarigal et al. 1995). Values express the high level of fragmentation and isolation of forest structures.

According to the Organisation for Economic Co-operation and Development (OECD, 2006), the Greater Milan area includes the provinces of Milan, Bergamo, Como, Lecco, Lodi, Monza and Brianza, Pavia, Varese, and the Piedmont province of Novara, amounting to a population of 7.4 million inhabitants and an area of 2944.53 km^2 (ISTAT, 2011). This makes it the largest Italian metropolitan region and among the top 10 in the European Union by population. The study area covers 1575 km^2, contains 134 municipal areas and totals over three million inhabitants (ISTAT, 2011) (Figure 1). The nominal GDP per capita amounts to €36 362, making it the richest province in Italy (UNIONCAMERE, 2010). The territory of the study area is located in central-west Lombardy and stretches from the Po River valley and Ticino River in the West to the Adda River in the East. It boasts a number of rivers (i.e., the Adda, Ticino, Olona, Lambro, and Seveso rivers) as well as a network of canals, including the Naviglio and Villoresi networks. The territory is also rich in groundwater and springs. The agricultural economy is highly productive but deeply incised by urban development.

With regard to climate, the Metropolitan Area of Milan can be described as a mesothermal sub-continental type, which is a transition between the Oceanic and Mediterranean climatic regions. The

Table 1. Spatial metrics describing the distribution or fragmentation of forest and use classes.

Province	Mean patch size (ha)	Patch density (n/Km²)	Nearest distance (m)	% patch > 15 ha
Mantova	1.89	0.45	1329	29.9
Lodi	2.88	1.1	757	29.17
Cremona	2.41	0.61	1109	21.32
Pavia	4.01	1.23	746	54.52
Brescia	7.17	0.7	889	75.89
Milano	4.33	1.48	589	59.45
Bergamo	9.39	1.23	555	84.32
Lecco	7.14	3.38	237	75.64
Como	15.93	2.08	550	90.05
Varese	23.41	1.58	248	92.71

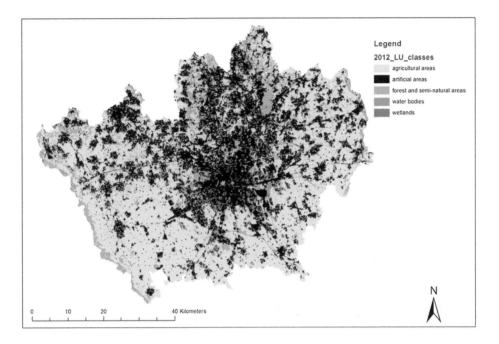

Figure 1. Distribution of land use classes in the Metropolitan Area of Milan (updated in 2012).

historical curves of total precipitation and average temperature for the period 1951–2006 (Sanesi, Lafortezza, Marziliano, Ragazzi, & Mariani, 2007) show a break-out point in the average temperature during the early 1990s, with an average increment of + 1.5 °C (from 12.5 to 14.5 °C). This break-out point follows the general climate pattern of the Euro-Mediterranean area (Werner, Gerstengarbe, Fraedrich, & Oesterle, 2000). Within this general trend, periods of drought and heat stress have occurred in recent years, especially during summer periods thus affecting the status of forest trees and vegetation in this area.

According to some authors (Battisti, 2006), the conditions of climate change currently affecting Italy and Lombardy not only encourage the spread of new diseases but also the expansion of areas of interest for exotic insect species. These insects cause severe damage and in some cases place the stability of forest ecosystems at risk.

In summary, the Metropolitan Area of Milan is represented by (i) high levels of urbanisation; (ii) high levels of soil sealing; (iii) a highly fragmented but still productive agricultural economy; and (iv) climate change impacts.

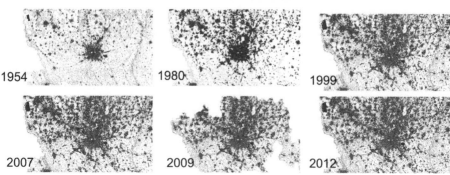

Figure 2. Urban expansion in the Metropolitan Area of Milan since 1954.

Materials and methods

We researched the dynamics of land-use change for a 60-year period starting in 1954 and applied them to UGI notably the Urban Forest component. This component includes discrete urban woodlands and landscapes that are strongly influenced by trees; these can be peri-urban or densely urban such as is the case with neighbourhood parks and boulevards. This investigation coincides with one of the most transformational periods of economic growth, social change and urbanisation in European history. Our analysis is based on work undertaken by the Region of Lombardy-ERSAF in the 1990s as part of the European Programme CORINE LAND COVER (CLC). Specifically, Lombardy Region-ERSAF created a tool for analysing and monitoring land use (DUSAF). This tool was shared through an 'Infrastructure for Spatial Information' (IIT) 'Geoportal' with other Italian regions. The data relate to land use for the years 1954, 1980, 1999, 2007, and 2012.

All information levels are comparable in that they use the same legend divided into three main levels, which are consistent with the specific CLC. The first of these includes five major categories of coverage, namely (i) artificial areas; (ii) agricultural areas; (iii) forest and semi-natural areas; (iv) wetlands; and (v) water bodies. These categories provide progressively detailed information at the second and third levels. This information in the database is consistent with a scale of 1:10 000 and consists of a polygonal component layer of land use and a linear component layer, which includes hedgerows and trees. We also took into account the main legislative interventions at different levels (regional, national and European), which have affected the agricultural economy and, in particular, those that deal with afforestation of agricultural land. Thus, we were able to highlight not only the changes in land use but also the effectiveness of the legislative measures in the course of 60 years. To highlight the dynamic economic and social context of land-use changes, we considered data from the Italian national statistical system for the main social and economic components. For this study, we determined that the analysis should only focus on the 'Province of Milan', now named Metropolitan area of Milan according to national law 56/2014. This would allow researchers to analyse more homogenous data based on recognisable administrative, social and economic parameters.

Results

Land-use changes

The changes in land use during the 1954–2012 time period show a high level of expansion in the urban built-up areas (i.e., sealed surfaces). In the study area from 1954 to 1980, this occurred notably at the expense of agricultural and forest land domains. During this time, sealed surfaces increased threefold from 23 740 to 75 012 hectares (Figure 2). This expansion rate of + 132% was followed by an additional expansion rate of + 24% from 1981 to 1999, which continued at a somewhat slower rate of + 8% in the

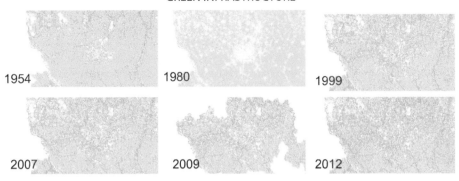

Figure 3. Loss of agricultural land in the Metropolitan Area of Milan.

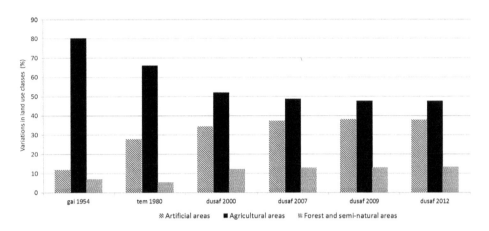

Figure 4. Percentage of variations in land use change from 1954 to 2012 compared to the total surface of the study area.

period between 2000 and 2007. However, in the past five years, which coincide with a serious economic crisis, the change in rate of urban expansion has been slight amounting to only 1%. In summary, there has been a continuous loss of agricultural land: 28 022 hectares from 1954 to 1980 (−18%), 27 468 hectares from 1981 to 1999 (−21%), 6870 hectares from 2000 to 2007 (−7%) and 2200 hectares in the period from 2008 to 2012 (−2%) (Figure 3).

The changes in forest areas (afforestation, forestry plantations, and secondary succession included) are, however, characterised by a different trend (Figure 4). After a loss of 3371 hectares in the 1954–1980 period (−24%), an increase of 13 600 hectares (+137%) was recorded in the subsequent period from 1981 to 1999 as well as in the periods 2000–2007 (+1241 hectares) and 2008–2012 (+790 hectares). In summary, the forest area increased by 12 260 hectares (+121%) in the period from 1954 to 2012. During the study period, the changes in wetlands and water bodies were minimal.

The National General Census of Agriculture for the last three decades clearly shows that during this time, the Metropolitan Area of Milan has undergone a process of progressive urbanisation and loss of the rural fabric in terms of both surface area and 'working' farms. The loss of agricultural land is reflected in the organisation of production. During the period from 1982 to 2010, the number of farms significantly decreased, while the average farm size increased. In 1982, there were 7133 farms with an average area of production (excluding forest land) of 10.90 hectares, whereas in 2010, there were only 2316 farms with an average area of production of 28.00 hectares. An outcome from the 1954–2012 study period is that soil consumption due to encroaching grey infrastructure (i.e., soil sealing) increased from 44.2 to 63.7% of the total regional soil surface (ISPRA, 2014). However, during the first half of this

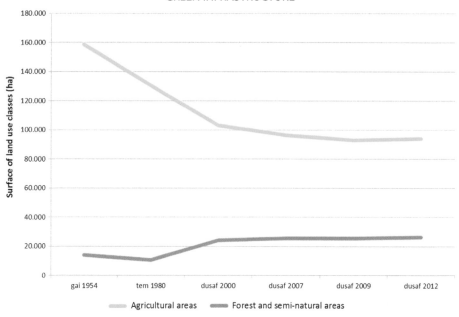

Figure 5. Trends in the extent of agricultural and forest land use in the study area from 1954 to 2012.

period (1954–1980), both agricultural and forest land decreased, whereas after 1980, the trend for the two domains diverged; agricultural land continued to decrease while forest land increased by a substantial amount (Figure 5).

Political and administrative measures

In recent decades, numerous administrative and legislative measures have influenced land-use changes in the Metropolitan Area of Milan. In Table 2, we report the key measures on a yearly basis. Also, a number of major initiatives have been carried out with the goal to preserve forest areas in the rural landscape and to create urban and peri-urban forests. The intention of these projects was to provide a range of services and products to people living in urban and sub-urban areas.

In the context of policy and legislative frameworks, it is important to note the results achieved with the adoption of the regional law (No. 86/1983) on protected areas. By 2012, the 'regional parks' system in Lombardy covered 22% of the land area and in the Province of Milan, this rises to 39%. The system consists of six regional parks aimed at retaining the nature of the landscape and certain habitats (e.g., Parco del Ticino) as well as the large agricultural belt south of Milan (Milan South Agricultural Park). The framework of protected areas also includes parks of local interest (PLIS). Currently, there are 14 PLIS covering 7000 hectares in the Metropolitan Area of Milan. Overall, the system of protected areas has extended to 48.6% of the total land area.

In 2014, the province of Milan (now Metropolitan Area of Milan) approved a Territorial Coordination Plan. This plan includes among its objectives (i) the protection of the environment, urban landscape, agricultural areas and the preservation of natural open spaces and (ii) the expansion and upgrading of the landscape system and environment with the construction of a network system of green spaces (so-called 'ecological network'). The rivers are important elements of this network system linking agricultural areas, forests and open spaces. There are also specific roles for such elements in 'Natura 2000' protected areas and in the creation of PLIS. There are a total of 17 PLIS in the metropolitan area covering a total area of 8,250 hectares and spread across 40 municipalities. The Plan also relies on a

Table 2. Yearly based description of political or administrative measures applied in the Milano Metropolitan Area

Time	Political/Administrative measures	Scope
1974	Municipal act for urban park 'Boscoincittà' establishment	Providing urban green areas (also in terms of forests) for citizens of western Metropolitan Area of Milan
1975	Regional law for urban park 'Parco Nord Milano' establishment	Providing urban green areas (also in terms of forests) for citizens of eastern Metropolitan Area of Milan
1983	Regional Law n.86/1983 of Protected Areas	Conservation and management of natural and environmental assets of the Lombardy region
1983	Regional law for urban park 'Oaks wood' establishment	Restoration and re-naturalisation of an area in Seveso (Milan) affected in 1976 by a chemical manufacturing (ICMESA) accident and the resulting pollution where the soil polluted by dioxin
1989	Regional Law n.80/1989	Measures for forest resources and for the protection of vegetation in the parks established by the Regional Law; initial projects of urban afforestation
1992	EEC Regulation 2080/1992	Forestry measures under the common agricultural policy; reforestation and afforestation of agricultural lands
1996	Regional Law n. 31/1996	First legislative recognition of the concept of 'green infrastructure' and its possible public financing because it operates in the public interest
2000	Regional Rural Development Plan 2000–2006	Action 2.6 establishment of hedgerows and action 2.8 afforestation of agricultural land
2000	Regional act 'Ten large forests for the plains'	Identification of 15 possible areas for afforestation, all equally distant from the centre of Milan
2001	Regional Law n. 221/2001	Introduction of the concept of compensatory afforestation linked to the transformation of the forest; the removing of forest woodland must be compensated through new afforestation projects.
2007	Rural Development Plan 2007–2013	Action 2.2.1 support afforestation, arboriculture planting, biomass, and Poplar on agricultural soils
2007	Regional act for 10.000 hectares of (new) forests; also called 'Green systems' act	In Metropolitan Area of Milan involves 17 projects
2009	Regional act for the establishment of the Regional Environmental Network	Promote the conservation of ecological resources through measures including ecological corridors connection between the priority areas for biodiversity, improving the quality of the habitat, the promotion of ecosystem functions.

strategic role of agricultural spaces and demonstrates that agriculture can play a key role in production, the conservation of biodiversity and water regulation.

New governance models

Whilst the principal reason why GI is being created and managed in the Metropolitan Area of Milan is for the ecosystem services, it is also recognised that it is a platform highly suited to new forms of governance. Policy-makers and municipal bodies are interested in new models of governance that are not solely based on direct land management by public bodies. There are a multiplicity of reasons behind this but one that is key in the prevailing economic climate is to lower the costs of land management by involving more citizens in direct management works. The history of some of the predominate UGI projects in the Metropolitan Area of Milan, such as 'Parco Nord Milano', has featured a strong participatory approach in terms of both promotion and planning since their inception. Elsewhere in the region of Lombardy, there are several experiences of urban forest parks built and managed through new models of governance and in these existing citizen associations have played a crucial role. The 'Bosco della Giretta' exemplifies a participative model from establishment through to management (Lawrence et al., 2013).

Discussion

There is currently a well-established range of protected agricultural areas, based on results of land-use change analysis, and an expanding urban and peri-urban forestry sector within the province of Milan. On a broader scale, however, these trends are in continual decline in overall agricultural land cover but increasing in forest land cover. Whilst the landscape scale structure of protected agricultural areas and the expanding urban and peri-urban forestry sector are complex, there is a discernable radial pattern centred on Milan, albeit with imperfections and notable gaps in connectivity. Discussion points are also nuanced by the impact of political and administrative measures on land-use change and by the robustness of how these are applied. It should be noted that whilst the terms 'GI' and 'UGI' have only recently been used in the Milan metropolitan area, many of the aspects of GI have been practiced in the region of Lombardy for some time, supporting the comments by Davies, McGloin, MacFarlane, and Roe (2006) that GI includes an element of 'old wine in new bottles'.

Political and administrative measures and their impact on land-use change

A comparison between land-use trends and protective administrative policies demonstrates a close connection between the two; furthermore, there are identifiable periods that characterise this relationship in terms of landscape scale change. These periods also reflect wider historical, social and economic transformations in the metropolitan area such as the post-WW2 economic revival, economic restructuring and liberalisation of the economy in the 1980s and 90s and, more recently, the recession of 2008 onwards.

The first period dates from 1954 to 1980. In the absence of measures for the protection of land, ecological and environmental resources, there was a dramatic loss of both agricultural and forestry lands in this period. The lack of an aerial survey has prevented us from identifying further sub-divisions in this timeframe. However, the entire economy of Milan grew during the period from 1954 until the early 1970s and it was only in the last part of this period that a decline commenced notably in the industrial sector. The 1954–1980 period was also characterised by high levels of immigration to meet the growing economic needs and this led to an expansion of the city and the development of new towns within its periphery. This development occurred in the context of weak spatial and development planning and the lack of protection of environmental resources and landscapes. By the end of the 1970s and in response to growing public interest, there was increasing concern in the Milan Municipality and regional administration of the Lombardy Region about the pace of land-use change, loss of habitats and the consequences to local communities. Planners and policy-makers responded by envisioning two great urban parks in the east and west of Milan. The result was 'Boscoincittà' and 'Parco Nord Milano', which were created on the basis of requests of citizens and environmental NGOs. The decision to afforest these parks was taken with the participation of the citizenry.

The second period is between 1981 and 1999. During this period, there was a continuing decline in agricultural land but also an inverse incremental increase in forest land that not only recovered the losses of the previous period but also led to a further increase. It is clear that agricultural and environmental policy measures have influenced this process. The crisis of the industrial sector also began the process of conversion of part of the industrial fabric into urban green areas such as the new urban Forlanini Park.

The third period covers the years 2000–2012. In this period, the loss of agricultural land is limited as there is less urban expansion. The forest resources had further increased through the creation of specific interventions such as the 'ten large forests for the plains' initiative. The economic crisis starting in 2008 is also reflected in the decrease of the population residing in the metropolitan area. However, during this same period, new demands arose from the citizens either through the expansion of social allotments, especially in the context of urban parks, and through the further spread of participatory processes. Towards the end of this period, 2012, the concept of GI is affirmed at the Provincial and Regional level.

Urban GI and urban forest evolution since 1980 in the Metropolitan Area of Milan

The establishment of new UGI and the protection of those previously developed have taken place due to the introduction of protective land management policies and the expansion of regional and local parks and urban forests. The process of afforestation of agricultural land has occurred within a remarkably short timeframe and this includes urban and peri-urban areas. This trend has been highlighted in the results of the European Union Life + Emonfur project.

Although urban forest resources do not constitute the majority of Milan's UGI, they do perform a vital role. In particular, woodland tree rows are the most visible product of policies to conserve, protect and guide management of land and they are also the main focus for publicly funded projects. Urban forests have been researched in some detail and the results show that they provide a range of ecosystem services (e.g., Ahern, 2011; Bowler, Buyung-Ali, Knight, & Pullin, 2010; Haase et al., 2014; Tzoulas et al., 2007) and this is specifically evidenced in the Metropolitan Area of Milan (Lafortezza, Carrus, Sanesi, & Davies, 2009; Marziliano, Lafortezza, Colangelo, Davies, & Sanesi, 2013; Mariani et al., 2016; Sanesi et al., 2007, 2009).

Presently (in 2015), the Lombardy Region is discussing a new regional law for the establishment of a REGIONAL METROPOLITAN PARK of Milan. In practice, this will lead to the creation of a greenbelt landscape scale park through the unification of current regional parks, notably Parco Nord Milano and South Milan Agricultural Park. They will be administered by a single operator with the intention that it will lead to improved efficiency and greater effectiveness in protecting and promoting the area. The initiative, in particular, aims to rationalise the management of green areas situated around the main urban environment. The single entity will be governed by public law and composed of the municipalities concerned and the metropolitan city of Milan.

Although there is an increased interest on the part of policy-makers and municipal bodies in citizen involvement for developing and planning UGI, experience in this context is still lacking and incomplete. This deficiency should be compensated by further efforts in participatory governance.

Conclusions

The Metropolitan Area of Milan in the Italian Region of Lombardy is a case study for how planning interventions and a legislative framework can affect the development and distribution of UGI. The researchers involved in this study have traced this development from 1954 until 2012 and are confident that the trends outlined are continuing. The European Union Life + project Emonfur has enabled an in-depth analysis of the urban forest domain in Milan. We have found that Urban Forests are especially important in the Metropolitan Area of Milan also because they provide an element that is perceived to be fundamental by city residents within accessible UGI. Projects such as Boscoincittà and Parco Nord Milano have become well-known international examples, as reflected by their hosting of the European Forum on Urban Forestry in 2013.

Between 1954 and the early/mid-1970s, there was a substantial decline in both agricultural and forest lands. However, since 1980, forest lands have increased rapidly whilst agricultural lands have continued to decline. This trend has recently slowed down, which we postulate is the result of the current economic difficulties faced across Europe.

It is clear from the Metropolitan Area of Milan that regulatory/legislative frameworks can have a significant impact on land-use changes. Project-based 'interventions' also have a considerable impact. The combination of the two has had a strongly positive result in terms of the regulatory role of ecosystem services in the province and to some extent is moderating urban development. Therefore, a lesson to be learned is that project-based interventions must necessarily be backed by regulatory/legislative frameworks to become effective tools.

In the Metropolitan Area of Milan, the landscape-scale pattern of UGI remains fragmented although a radial pattern of UGI is discernable. However, green areas such as Urban Forests are accessible to residents albeit this can involve the use of public transport to a destination rather than on foot. New urban forest areas such as Parco Nord Milano are however still immature landscapes and the benefits they provide will continue to accumulate for many years to come. The establishment of a REGIONAL METROPOLITAN PARK of Milan through the union and synergy of the major existing parks can significantly contribute to the consolidation of UGI. The consolidation process of UGI cannot take place solely through the direct intervention of public institutions.

There is current interest in new forms of governance in relation to UGI in the Metropolitan Area of Milan. We believe that this reflects a desire by planners and policy-makers to lighten the burden of maintenance of green spaces and to involve citizens in such activities. There are still few experiences in which the involvement of citizens was focused in all management processes and phases. In the future, the process of local citizen participation would benefit by referring to similar experiences that are well underway in other European contexts.

Geographical terms adopted in this paper and their extent

In the hierarchical organisation of the Italian Republic, the central state defines the legal framework which each region refers to in its own legislation. The state is divided in several building blocks that are (in administrative scale) regions, provinces, municipalities and metropolitan cities. There are 20 regions; each region is subdivided into provinces, which, at the level of large urbanised areas, are called metropolitan cities. Provinces and metropolitan cities are further organised into municipalities. Milan is a metropolitan city.

Disclosure statement

No potential conflict of interest was reported by the authors.

ORCID

Giovanni Sanesi http://orcid.org/0000-0002-4218-3605
Giuseppe Colangelo http://orcid.org/0000-0002-1124-4399

References

Ahern, J. (2011). From fail-safe to safe-to-fail: Sustainability and resilience in the new urban world. *Landscape and Urban Planning, 100*, 341–343.

Barbati, A., Corona, P., Salvati, L., & Gasparella, L. (2013). Natural forest expansion into suburban countryside: Gained ground for a green infrastructure? *Urban Forestry and Urban Greening, 12*, 36–43.

Battisti, A. (2006). Insect populations in relation to environmental change in forests of temperate Europe. In: T. D. Paine, (Ed.), *Invasive forest insects, introduced forest trees, and altered ecosystems: Ecological pest management in global forests of a changing world* (pp. 127–140). The Netherlands: Kluwer Academic Publishers.

Benedict, M. A., & McMahon, E. T. (2002). Green infrastructure: Smart conservation for the 21st century. *Renewable Resources Journal, 20*, 12–17.

Blotevogel, H. H. (1998). The Rhine-Ruhr metropolitan region: Reality and discourse. *European Planning Studies, 6*, 395–410.

Bowler, D. E., Buyung-Ali, L., Knight, T. M., & Pullin, A. S. (2010). Urban greening to cool towns and cities: A systematic review of the empirical evidence. *Landscape and Urban Planning, 97*, 147–155.

Cassatella, C. (2013). The 'Corona Verde' strategic plan: An integrated vision for protecting and enhancing the natural and cultural heritage. *Urban Research & Practice, 6*, 219–228.

Collingham, Y. C., & Huntley, B. (2000). Impacts of habitat fragmentation and patch size upon migration rates 2000. *Ecological Applications, 10*, 131–144.

Davies, C., McGloin, C., MacFarlane, R., & Roe, M. (2006). *Green infrastructure planning guide project: Final report.*

Davies, C., Hansen, R., Rall, E., Pauleit, S., Lafortezza, R., DeBellis, Y., Santos, A., & Tosics, I. (2015). *Green Infrastructure Planning and Implementation (GREEN SURGE). The status of European green space planning and implementation based on an analysis of selected European city-regions* (Report No. 5.1). GREEN SURGE, FP7-ENV.2013.6.2-5-603567; 2013-2017. 1–134.

European Environment Agency. (2006). *Urban sprawl in Europe – The ignored challenge* (EEA Report No. 10/2006). Copenhagen: European Environmental Agency (EEA).

European Union. (2014). *Eurostat regional yearbook.* Luxembourg: Publications Office of the European Union.

Forman, R. T. T. (1995). *Land mosaics. The ecology of landscapes and regions.* New York, NY: Cambridge University Press.

Geneletti, D. (2004). Using spatial indicators and value functions to assess ecosystem fragmentation caused by linear infrastructures. *International Journal of Applied Earth Observation and Geoinformation, 5*(1), 1–15.

Haase, D., Larondelle, N., Andersson, E., Artmann, M., Borgström, S., Breuste, J., … Elmqvist, T. (2014). A quantitative review of urban ecosystem service assessments: Concepts, models, and implementation. *Ambio, 43*, 413–433.

Hansen, R., & Pauleit, S. (2014). From multifunctionality to multiple ecosystem services? A conceptual framework for multifunctionality in green infrastructure planning for urban areas. *AMBIO: A Journal of the Human Environment, 43*, 516–529. doi:10.1007/s13280-014-0510-2.

ISPRA. (2014). *Il consumo di suolo* [The soil loss in Italy].

ISTAT. 2011. *Superficie dei comuni, province e regioni al Censimento 2011* [Muncipalities, provinces and regions surfaces on 2011 general census].

Kasanko, M., Barredo, J. I., Lavalle, C., McCormick, N., Demicheli, L., Sagris, V., & Brezger, A. (2006). Are European cities becoming dispersed?: A comparative analysis of 15 European urban areas. *Landscape and Urban Planning, 77*, 111–130.

Lafortezza, R., Carrus, G., Sanesi, G., & Davies, C. (2009). Benefits and well-being perceived by people visiting green spaces in periods of heat stress. *Urban Forestry & Urban Greening, 8*, 97–108.

Lafortezza, R., Davies, C., Sanesi, G., & Konijnendijk, C.C. (2013). Green infrastructure as a tool to support spatial planning in European urban regions. *IForest, 6*, 102–108.

Lawrence, A., De Vreese, R., Johnston, M., Konijnendijk, C. C., & Sanesi, G. (2013). Urban forest governance: Towards a framework for comparing approaches. *Urban Forestry & Urban Greening, 12*, 464–473.

Llausàs, A., & Roe, M. (2012). Green infrastructure planning: Cross-national analysis between the North East of England (UK) and Catalonia (Spain). *European Planning Studies, 20*, 641–663.

Lovell, S. T., & Taylor, J. R. (2013). Supplying urban ecosystem services through multifunctional green infrastructure in the United States. *Landscape Ecology, 28*, 1447–1463.

Madureira, H., & Andresen, T. (2014). Planning for multifunctional urban green infrastructures: Promises and challenges. *Urban Design International, 19*, 38–49.

Madureira, H., Andresen, T., & Monteiro, A. (2011). Green structure and planning evolution in Porto. *Urban Forestry and Urban Greening, 10*, 141–149.

Marziliano, P. A., Lafortezza, R., Colangelo, G., Davies, C., & Sanesi, G. (2013). Structural diversity and height growth models in urban forest plantations: A case-study in northern Italy. *Urban Forestry & Urban Greening, 12*, 246–254.

Mariani, L., Parisi S. G., Cola, G., Lafortezza, R., Colangelo, G. & Sanesi, G. (2016). Climatological analysis of the mitigating effect of vegetation on the urban heat island of Milan, Italy. *Science of the Total Environment, 569–570*, 762–773.

McGarigal, K., & Marks, B. J. (1995). *FRAGSTATS: Spatial pattern analysis program for quantifying landscape structure*. Gen. Tech. Rep. PNW-GTR-351. Portland, OR: U.S. Department of Agriculture, Forest Service, Pacific Northwest Research Station. 122 p.

Mell, I. C. (2009). Can green infrastructure promote urban sustainability? *Proceedings of the Institution of Civil Engineers: Engineering Sustainability, 162*, 23–34.

Mell, I. C. (2010). *Green Infrastructure: Concepts, perceptions and its use in spatial planning* (Unpublished doctoral dissertation). Newcastle University, Newcastle, UK.

OECD. (2006). OECD territorial reviews. Milan, Italy.

ONCS. (2009). *Primo rapporto nazionale sui consumi di suolo* [First report on soil loss]. Osservatorio Nazionale sul Consumo di Suolo (ONCS), Maggioli Editore.

Roe, M., & Mell, I. (2013). Negotiating value and priorities: Evaluating the demands of green infrastructure development. *Journal of Environmental Planning and Management, 56*, 650–673.

Sanesi, G., Lafortezza, R., Marziliano, P. A., Ragazzi, A., & Mariani, L. (2007). Assessing the current status of urban forest resources in the context of Parco Nord, Milan, Italy. *Landscape and Ecological Engineering, 3*, 187–198.

Sanesi, G., Padoa-Schioppa, E., Lorusso, L., Bottoni, L., & Lafortezza, R. (2009). Avian ecological diversity as an indicator of urban forest functionality. Results from two case studies in northern and southern Italy. *Arboriculture & Urban Forestry, 35*, 80–86.

Tzoulas, K., Korpela, K., Venn, S., Yli-Pelkonen, V., Kaźmierczak, A., Niemela, J., & James, P. (2007). Promoting ecosystem and human health in urban areas using green infrastructure: A literature review. *Landscape and Urban Planning, 81*, 167–178.

UN. (2014). World urbanization prospects by UN DESA's population division.

UNIONCAMERE. (2010). *L'economia reale dal punto di osservazione delle Camere di Commercio* [The real economy from the vantage point of Chambers of Commerce]. Rome, Italy.

Weber, T., Sloan, A., & Wolf, J. (2006). Maryland's green infrastructure assessment: Development of a comprehensive approach to land conservation. *Landscape and Urban Planning, 77*, 94–110.

Werner, P. C., Gerstengarbe, F. W., Fraedrich, K., & Oesterle, K. (2000). Recent climate change in the North Atlantic/European sector. *International Journal of Climatology, 20*, 463–471.

Young, R. F. (2011). Planting the living city. *Journal of the American Planning Association, 77*, 368–381.

Young, R. F., & McPherson, E. G. (2013). Governing metropolitan green infrastructure in the United States. *Landscape and Urban Planning, 109*, 67–75.

Can we face the challenge: how to implement a theoretical concept of green infrastructure into planning practice? Warsaw case study

Barbara Szulczewska, Renata Giedych and Gabriela Maksymiuk

ABSTRACT
While green infrastructure (GI) is now a widely referenced concept, in Poland it is still only discussed among academics and has yet to be implemented. In 2013, the EC recommended that Member States promote the implementation of GI approaches. In recent decades, ecological discourse has dominated Polish cities' planning practice. In Warsaw, this discourse has evolved into the formation of the Warsaw Natural System (WNS) concept. This study examines means of transforming the WNS, which is strictly related to ecological discourse, into a Warsaw Green Infrastructure (WGI). In doing so, we utilise basic principles of GI, namely: integration, multifunctionality, connectivity, and multi-scale and multi-object approaches. The authors expose the main challenges associated with utilising the WNS concept as a point of departure for WGI implementation.

1. Introduction

In recent years, several planning concepts to maintain and enhance green spaces in cities have been applied. Such approaches include the following: ecological networks (Bryant, 2006; Ignatieva, Stewart, & Meurk, 2011; Jongman, Külvik, & Kristiansen, 2004; Opdam, Steingröver, & Rooij, 2006), ecological land use complementation (Colding, 2007; Goddard, Dougill, & Benton, 2010; Jones, Howard, Olwig, Primdahl, & Herlin, 2007), conservation subdivision (Arendt, 2004; Carter, 2009; Freeman & Bell, 2011), low impact development (Dietz, 2007; Pyke et al., 2011) and green infrastructure (GI) (Benedict & McMahon, 2006; Hostetler, Allen, & Meurk, 2011; Mell, Henneberry, Hehl-Lange, & Keskin, 2013 Mell, 2014). Recently, the GI concept has attracted the most interest. Hundreds of scientific publications, guidelines, recommendations and evaluations outlining more or less detailed conceptualisations and definitions of GI have been published worldwide. The scope of these publications has varied depending on the scale and the areas examined. Generally speaking, the following—albeit overlapping—approaches to GI conceptualisation can be identified:

(1) The 'structural' approach (based on ecological network theories), which focuses on the protection of ecosystems and related services (Benedict & McMahon, 2006; Hostetler et al., 2011; Ignatieva et al., 2011);
(2) The 'hydrological' approach, which emphasises sustainable water management practices (Ahern, 2007; Dietz, 2007; Pyke et al., 2011);

(3)   The 'integrated' approach, which particularly highlights the need to integrate various functions, from nature conservation to social benefits for citizens (Madureira, Andresen, & Monteiro, 2011; Mell et al., 2013; Niemelä et al., 2010).

These three approaches to study GI are legible in planning practice instruments, when the GI concept is actualised through GI plans. The structural approach is more often associated with a regional scale, where connectivity between 'hubs' and 'corridors' is stressed, as exemplified in the Maryland's Green Infrastructure Assessment and GreenPrint Program (2004). Among the cities where the 'hydrological' approach was a basis for GI formation are American cities, such as New York, with its NYC Green Infrastructure Plan, A Sustainable Strategy for Clean Waterways (2010, NYC Green Infrastructure Plan, 2014 Annual Report, 2014) and Chicago and Lancaster (City of Chicago Green Stormwater Infrastructure Strategy, 2014; Green Infrastructure Plan, City of Lancaster, 2011). On the other hand, a more 'integrated approach' is presented in Barcelona (Barcelona Green Infrastructure and Biodiversity Plan, 2013), Vitoria-Gasteiz (The Interior Green Belt: Towards an urban green infrastructure in Vitoria-Gasteiz, 2012) and London (Green Infrastructure & Open Environments: The All London Green Grid, 2012).

Despite the popularity of the GI approach, the concept is not widely discussed among professionals in some European countries, including Poland (Giedych, Szulczewska, Dobson, Halounova, & Doygun, 2014). Nevertheless, given the potential role of GI in shaping urban environments through the delivery of ecological services, it must be recognised as an important urban planning instrument and as a strategic approach to sustainable urban development that combines land conservation and land use planning (Benedict & McMahon, 2006; Hostetler et al., 2011; Madureira et al., 2011; Novotny, Ahern, & Brown, 2010).

The European Commission (2013) sets forth the reasons and the need for the implementation of the concept within urban areas as well. It could be considered as a driving force for the development of GI strategy in European cities, as well as in countries where the GI concept has not yet been adopted.

We thus explore and evaluate strengths and limitations of GI implementation in Polish conditions in consideration of existing concepts that have evolved in Poland and that are characteristic for our spatial planning tradition. In this paper, we present and discuss our proposal to implement the GI idea, taking as the point of departure the already existing concept of the Warsaw Natural System (WNS)—a concept aimed at maintaining the correct proportion between built-up and open areas, which is strictly related to ecological discourse and firmly rooted in the Warsaw spatial planning tradition. At the same time, we discuss possibilities and limitations of the implementation of GI principles presented in publications into practice.

2. Materials and methods

2.1. The study area

Warsaw is perceived as a green city. Taking into account the available data (Statistical Office in Warsaw, 2015) on land use structure, land uses, which may constitute the main body of GI, account for 51% of the total Warsaw area (Figure 1). The dominant present land use within those elements is agriculture, and therefore arable lands, pastures, meadows and orchards cover almost 22% of total city area. The second biggest land use category is forests (17% of total city surface), and they are located mainly in urban peripheries. Green spaces represented by 88 parks, 233 smaller pocket parks, allotment gardens, cemeteries, estate and street greenery cover an area of 4 572 ha (8.8% of city area). However, despite a rather large area, the distribution of green spaces in the urban fabric is uneven, as peripheral districts lack public parks. The green space structure is shaped as a combination of a band and wedge system that reflects the natural Warsaw relief with its Vistula river valley, the Warsaw Escarpment and a moraine plateau. Searching for potential for the GI formation in Warsaw, one should consider the outstanding qualities of these elements:

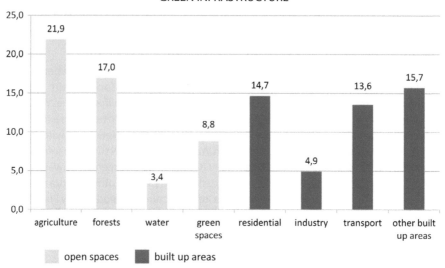

Figure 1. Warsaw land-use structure (in per cents). Source: Statistical Office in Warsaw, 2015.

(1) Natural character of a wide river valley, with features characteristic of a braided river, such as sandbars, floodplains and riparian forests;

(2) A significant landmark of the Warsaw Escarpment whose elevation varies from approximately 20 m in the Old Town area and central district to 10 m in the northern and southern quarters together with forests, magnificent historical parks and gardens, sporting and recreational areas, palaces and other historical buildings that are found along the escarpment;

(3) Still remaining natural landscape, such as patches of forest situated primarily in the outskirts and main alleys of trees that cross the city centre.

2.2. Warsaw Natural System

The concept of the WNS was introduced into Warsaw's Spatial Policy (Warsaw Environmental Study, 2006, amended 2010, 2014), on the basis of the Warsaw Environmental Study (2006) following the theoretical concept published by Szulczewska and Kaftan (1996). It can be defined as a part of the city's area dedicated to stabilising and enhancing environmental processes in the whole city. In order to play this role, areas which belong to the WNS must be developed with particular consideration.

The WNS approach as outlined in the Warsaw Environmental Study (2006) aimed to identify territorial units (areas) relevant for the climatic, hydrological and biological performance of the city. Finally, the WNS was then delineated via subsystems (climatic, hydrological and biological) overlapping analysis. If all three subsystems overlapped, they formed the main body of the WNS, or the so-called core area. Areas that were critical to the performance of two (typically hydrological and biological) subsystems were deemed supporting areas. Areas that heavily influenced the climatic subsystem were also considered as supporting areas.

The final version of the WNS (Figure 2) that was introduced in the Warsaw Spatial Policy includes both major structural elements (core and supporting areas) and linkages between WNS areas (green spaces and street greenery, which form main ecological connections).

The provisions of the Warsaw Spatial Policy in relation to the WNS define its borders, acceptable land use types (Figure 3) and Ratio of Biologically Vital Area (RBVA)—measurements which determine the amount of area covered by vegetation and/or water bodies in relation to sealed surfaces of the spatial unit (Szulczewska et al., 2014).

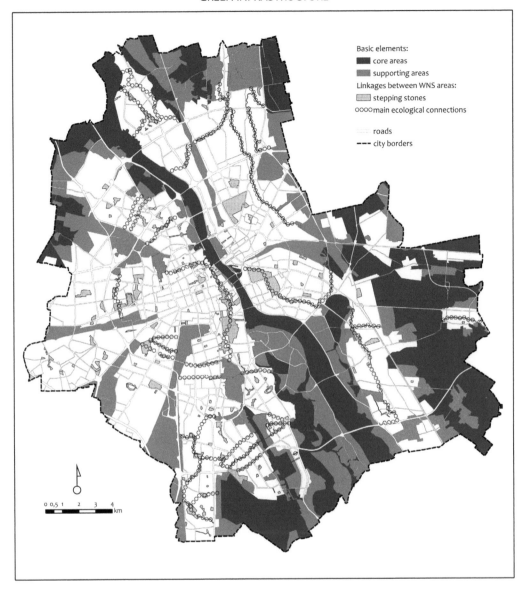

Figure 2. WNS. Source: Elaborated on the basis of Warsaw Spatial Policy, 2006, amended 2010, 2014.

2.3. Adopted methodological approach

2.3.1. Main assumptions

In addressing the possibility of the implementation of the GI concept in Warsaw, we decided to concentrate on the principles that Hansen and Pauleit (2014) identified on the basis of Benedict and McMahon (2006), Kambites and Owen (2006), and Pauleit, Liu, Ahern, and Kazimierczak (2011). Originally, they listed two groups of approaches: addressing the green structure and addressing the governance process. However, as our focus is an examination of the possibilities of embedding GI into the existing Warsaw Spatial Policy, we limited our investigation only to the principles addressing the spatial green structure: (1) integration, (2) multifunctionality, (3) connectivity, (4) a multi-scale approach and (5) a multi-objects approach.

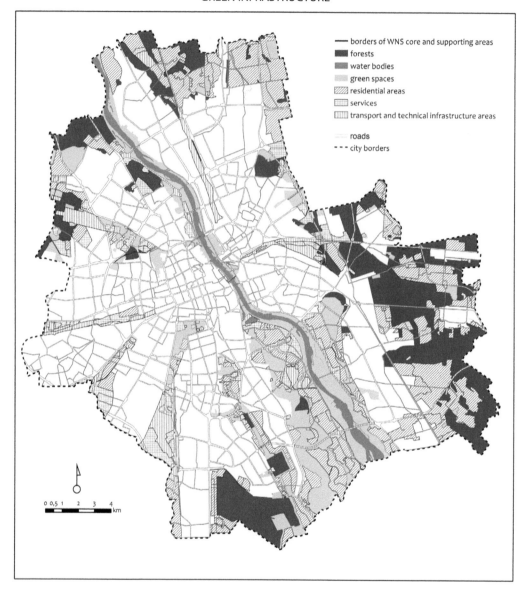

Figure 3. WNS planned land–use structure. Source: Elaborated on the basis of Warsaw Spatial Policy, 2006, amended 2010, 2014.

For the purposes of this study, we define Warsaw Green Infrastructure (WGI) as a system of functionally and/or structurally interconnected objects that are covered by vegetation and water, and which are considered important for a city's environmental performance, climate change adaptation and residents' quality of life.

2.3.2. Materials

Our primary research materials were spatial planning documents: the Warsaw Spatial Policy (Warsaw Spatial Development Conditions and Directions Study 2006 with the amendments of 2010 and 2014) and the local plans (235). The Spatial Policy is a mandatory planning document at the municipal level in Poland, which provides a system of principles to guide decisions on land use and the spatial structure of the city. It also endorses WNS—a concept considered to be relevant for the implementation of WGI (see point 2.2). According to Polish law on spatial planning (Spatial Planning and Development Act, 2003), legally, this policy is a statement of intent, and it is implemented by the local plans (acts of local spatial law).

In our analysis, we utilised publications which refer to GI implementation strategies (with special consideration given to those describing success stories) in order to establish the scope of the policies that are relevant for GI.

2.3.3. Analyses: scope and procedures

A content analysis of the Warsaw Spatial Policy and the local plans enabled us to discuss the possibility and conditions for the implementation of GI in the updated version of this Policy, which is supposed to be elaborated in subsequent years.

According to the main principles of GI presented by Hansen and Pauleit (2014), we conducted five analyses aimed at recognising WNS characteristics on the background of those principles and their interpretation available in the literature. Based on theoretical considerations and guidelines, in each case we formulated a set of questions derived from particular principle description. Then, we responded to those questions taking into account provisions of the Warsaw Spatial Policy related to WNS and relevant provisions of other policy sectors, which may or should be important for transforming the WNS concept into the WGI model.

In surveying the *integration principle*, we have tried to answer the questions related to the policy sectors relevant for the implementation of WNS and the background of the policy sectors pertinent to GI. Hansen and Pauleit's (2014) integration approach was the first to address the green structure concept, and it has been endorsed in many guidelines and strategies on GI implementation. They indicated the need to integrate GI planning with planning other urban infrastructure (e.g. transport, water management) and structures (e.g. built-up areas). In order to follow this guideline, policy sectors that are considered relevant to GI implementation must be identified and utilised.

Table 1. Policy sectors determined as necessary for GI implementation.

Policy sector to be included in GI planning and management	Analysed guides and strategies							
	1	2	3	4	5	6	7	8
Development policy		x		x	x	x	x	
Spatial policy		x		x				
Housing policy		x			x			
Transportation		x				x		
Food production		x					x	
Energy production/efficiency		x						
Rainwater, flooding and wastewater management	x	x	x		x	x	x	x
Climate change mitigation/adaptation				x	x		x	x
Nature protection (biodiversity)					x			x
Cultural heritage protection						x		x
Landscape protection		x						x
Fibre and fuel production				x				x
Health		x			x			x
Education							x	x
Recreation				x	x	x	x	
Public space							x	

(1) Rutherford (2007) The GI Guide. Issues, Implementation Strategies and Success Stories
(2) Natural England (2008) The Essential Role of GI: Eco-towns GI Worksheet
(3) American Rivers, Association of State and Interstate Water Pollution Control Administrators, National Association of Clean Water Agencies, Natural Resources Defence Council, The Low Impact Development Centre, U.S. Environmental Protection Agency (2008) Managing Wet Weather with GI Action Strategy
(4) Land Use Consultants (2009) South East GI Framework. From Policy into Practice
(5) Natural Economy Northwest (2009) A Guide to Planning GI at the Sub-Regional Level DRAFT (v3.1)
(6) Green Infrastructure Centre and E² Inc. (2010) Richmond Green Infrastructure Assessment
(7) Landscape Institute (2011) Local GI. Helping communities make the most of their landscape
(8) The Royal Institution of Chartered Surveyors (2011) GI in Urban Areas. RICS Practice Standards

Analysis of eight different GI implementation guides and strategies (Table 1) revealed that depending on the document, varied sets of policies have been recommended. It is obvious that discrepancies in the examined approaches have resulted from differing conceptualisations of the 'GI' term. Additionally, the analysis revealed that water and wastewater management should guide the GI implementation policy sector. This view, however, results from a very strong conceptualisation of GI as a measure for promoting sustainable water (particularly storm water) management (Ahern, 2007; Dietz, 2007; Pyke et al., 2011). All of these findings demonstrate that during GI implementation, relevant policy sectors must be identified independently, according to specific problems of each city.

In examining the *multifunctionality principle*, we have tried to compare the scope of the functions assigned for WNS with the GI functions and to explore the possibility of the implementation of this principle, while formulating the Warsaw Spatial Policy's provisions. Hansen and Pauleit (2014) explain that *multifunctionality principle* aims at intertwining or combining different functions, and thus using limited space more effectively. Taking into account numerous, and sometimes conflicting functions of GI (Green Infrastructure and Territorial Cohesion, 2011) it seems that this principle can be fully implemented at the site level for particular GI elements. Concepts and frameworks for GI multifunctionality assessment and planning presented by Hansen and Pauleit (2014) revealed that the whole procedure seemed to be rather demanding. Utilisation of the ecosystem services concept in this procedure requires a proper set of indicators, whose elaboration demands data rarely available for all potentially considered services.

In considering the *connectivity principle*, we were interested in the extent to which WNS was planned as a coherent structure and whether WGI could follow this structure. Hansen and Pauleit (2014) defined connectivity as physical and functional connections between green spaces. This connectivity, according to e.g. Ahern (2007) and Chang, Li, Huang and Wu (2012), is usually referred to as mainly ecological. Hansen and Pauleit (2014) point out that in the case of GI, the term connectivity should be understood in a wider context and linked to its multiple functions. They also refer to Davies, MacFarlane, McGloin, and Roe (2006) and their matrix (linking quality of GI elements with the connectivity). Connectivity analyses also require addressing two crucial issues: (1) Establishment and maintenance of corridors (Fleury & Brown, 1997; Osborn & Parker, 2003; Palmores, 2001; Viles & Rosier, 2001) or greenways (Ahern, 1995, p. 2) Presence of barriers (physical obstructions) or gaps (interruptions to continuity) (Parker, Head, Chisholm, & Feneley, 2008).

In analysing the *multi-scale principle*, we were interested in the extent to which the implementation of WNS has followed this principle and whether the rules applied for WNS could be utilised for WGI. Hansen and Pauleit (2014) mentioned that GI planning was to be performed at different scales: from the individual parcel to the state. This multi-scale approach is typically applied in Europe for the purposes of GI planning and GI spatial analysis (European Environment Agency, 2014). This approach is also supported in extensive studies on ecological networks (e.g. Bryant, 2006; Ignatieva et al., 2011; Jongman et al., 2004; Opdam et al., 2006).

In the case of the *multi-object principle*, we have identified the types of land uses planned for WNS that can be recognised as WGI elements. Hansen and Pauleit (2014) emphasised that different kinds of urban green and blue can be considered as GI objects. According to Benedict and McMahon (2006), GI consists of a wide range of open spaces that maintain ecological processes, e.g. waterways, wetlands, forests, wildlife habitats and greenways. In urban areas, GI may also refer to green roofs and vegetated bioswales (Ahern, 2007), domestic gardens (Goddard et al., 2010) and street trees (Ignatieva et al., 2011). Davies et al. (2006) present a wide selection of GI elements in a detailed typology. Represented overall categories are: agriculture (including 11 GI elements, e.g. horticulture, stock grazing), green spaces (including nine different GI elements, e.g. public parks and gardens, public provision for children and young people), transportation (with greenways, quiet lanes, cycle routes or canals), burial grounds (e.g. cemeteries, disused churchyards), restricted access green spaces (e.g. retail park settings) or controlled access green spaces (e.g. airports and military training land), vacant land (e.g. land identified for development or derelict land), waterways and water reservoirs (e.g. rivers, ponds, wetlands) and other open spaces such as beaches or dunes. From this wide selection of GI objects, one can suppose that any object or area covered by vegetation and/or water should be considered as a GI element. The problem appears that different types of GI objects should be considered at different levels of GI implementation and from this point of view the multi-scale approach seems to be crucial.

3. Results: questions and answers for Warsaw

3.1. Integration principle

- Which policy was considered as the most appropriate to implement the WNS? Is it correct from the WGI implementation point of view?

Table 2. Integration of WNS implementation provisions with the provisions of other policies.

Spatial policy sectors relevant to GI development (based on Table 1)	Policy sector provisions that may affect WNS development	Provisions directly targeted at WNS development
Development policy	This policy does not contain provisions related to GI	
Spatial policy	Minimum ratio of biologically vital areas (25–40% for residential areas)	Minimum ratio of biologically vital areas (40–60% for residential areas located within the WNS) Minimum ratio of biologically vital areas (70–90% for green spaces)
Housing policy	This policy does not contain provisions related to GI	
Transport	Bicycle route system developed as a supporting transportation system	No specific provisions
Food production	All existing agricultural areas are to be transformed into residential or green spaces (parks or sports centres)	No specific provisions, though a significant number of existing agricultural areas are located within the WNS framework
Energy production/efficiency	This policy does not contain provisions related to GI	
Rainwater, flooding and wastewater management	Recommended sustainable drainage system and rainwater management along the borders of individual plots Prohibition/ban on building within 10 m of watercourses and water bodies	No specific provisions
Climate change mitigation/ adaptation	Not applicable; this policy was not included in Warsaw Spatial Policy	
Nature protection (biodiversity)	Reference to existing and planned nature protection areas and guidelines on their protection	The WNS is deemed a structure that is responsible for the environmental performance of the city of Warsaw (other functions e.g. residential, recreational are to be subordinate to the main function) Special guidelines on protection areas are included in the WNS (among others: green connections—a minimal width of 10 m)
Cultural heritage protection	Reference to existing historical heritage conservation zones and to guidelines on their protection/development Designation of areas that are to be protected as cultural parks	No specific provisions, though some protected objects or objects to be protected are located within the WNS framework
Landscape protection	Reference to natural relief and anthropogenic forms that are deemed important landmarks	No specific provisions, though some landmarks are located within the WNS framework
Fibre and fuel production	Not applicable; this policy was not included in Warsaw Spatial Policy	
Health	This policy does not contain provisions related to GI	
Education	This policy does not contain provisions related to GI	
Recreation	Standards: parks—10 m^2 per inhabitant sports areas—40 m^2 per inhabitant, walking distance to green spaces—10–20 min	No specific provisions
Public space	The enhancement, development and redevelopment of public spaces with particular attention to social and aesthetic attributes	No specific provisions

Provisions of the Warsaw Spatial Policy, directly related to the WNS, were placed in the section dedicated to environmental protection policy. That is in tune with the main assumption on which the WNS concept was based: to safeguard and protect space responsible for the environmental performance of the city. In the case of the WGI, however, such placement seems to be inappropriate. Taking into account the general GI idea, its role and functions, as well as the wide selection of policies important for GI development and implementation, three options of WGI presentation in planning documents may be considered: (1) WGI as an 'independent' policy sector on the same level as e.g. transportation or housing policy; (2) WGI as part of the chapter on spatial structure of the city; (3) WGI as an element of different policies already included into the Warsaw Spatial Policy, but upgraded in terms of WGI implementation.

- To what extent provisions of policy sectors, already included into the Warsaw Spatial Policy, may be important for WGI implementation?

First, it should be noted that most of the policies (Table 2) already encompassed—as policy sectors—in the Warsaw Spatial Policy were considered to be relevant for the implementation of GI. Of course, the scope of those policy sector provisions should be expanded, as they were designed only for WNS, and not in order to introduce a more multifunctional WGI.

Table 2 also shows that not all of the policy sectors considered to be relevant for the implementation of GI (such as development policy, health, housing, education, fibre and fuel production) are incorporated in the Warsaw Spatial Policy. This means that the WGI concept should be included, while other strategic documents are to be prepared (usually in the case of Polish municipalities: a development strategy, an environmental protection programme and a climate change adaptation/mitigation policy, which have just started to be elaborated in Polish cities).

3.2. Multifunctionality principle

- What sort of functions are intended to be developed within the WNS, according to the Warsaw Spatial Policy?

According to the WNS concept (see Section 2.2), environmental functions (climatic, hydrological, biological) must be considered as the principal ones. As stated in the Warsaw Spatial Policy, subordinated to them are other functions, such as: residential, recreational, leisure and aesthetical. It should be noted that those functions were mentioned as 'among others', which means that it is an open list. The reason for such an approach lies in the WNS concept, where not function but location and environmental characteristics of the area are crucial for its incorporation into the system.

- To what extent are functions planned within the WNS consistent with GI functions?

In order to refer to the multifunctionality issue, we could only compare the main assumptions for WNS and WGI creation and identify planned functions of the areas included in the WNS, taking into account planned land uses. That allowed us to identify functions, however it did not permit a full evaluation of their scope and meaning for the particular land use as well as for the whole GI network. The point of departure for function identification was their enumeration included in the publication: Green Infrastructure and Territorial Cohesion (2011). Results of the analysis are presented in Table 3.

Analysis of Table 3 revealed that not all GI functions were planned to be developed within the WNS. It is obvious that due to the general WNS idea and its character, first of all environmental functions were planned. Further study of the provisions of the Warsaw Spatial Policy sectors (e.g. spatial structure, cultural heritage protection, water management), referring to issues other than WNS, allowed us to identify that some social functions were also assigned to selected areas within the types of land use included in the WNS. There are, however, GI functions that are not mentioned in the Warsaw Spatial Policy: e.g. maintaining the potential for agricultural land and providing space for renewable energy. This occurred because the Warsaw Spatial Policy's provisions did not foresee the types of land uses which could host those functions.

Table 3. WGI functions that are planned to be developed within the WNS.

GI functions according to Green Infrastructure and Territorial Cohesion (2011)	Parks	Green spaces adjacent to sport and leisure facilities	Historic parks in forests	Forests	Vistula river green spaces	Cemeteries	Built-up areas (min. 60% of rbva)	Single-family housing	Single-family housing in forests	Health care services	Sport and recreation areas	Educational services
Habitats for species	+	+	+	+	+	+	+	+	+	+	+	+
Permeability for migrating species	+	+	+	+	+	+	+	+	+	+	+	+
Connecting habitats	+	+	+	+	+	+	+	+	+	+	+	+
Mitigating urban heat island effect; corridors for air ventilation	o	o	o	o	o	o	o	o	o	o	o	o
Storing flood water; ameliorating surface water run-off	o	o	o	o	o	o	o	o	o	o	o	o
Carbon sequestration	–	–	–	–	–	–	–	–	–	–	–	–
Encouraging sustainable travel	–	–	–	–	–	–	–	–	–	–	–	–
Reducing energy use (heating and cooling)	–	–	–	–	–	–	–	–	–	–	–	–
Providing space for renewable energy	–	–	–	–	–	–	–	–	–	–	–	–
Sustainable drainage systems attenuating surface water run-off	o	o	o	o	o	o	o	o	o	o	o	o
Groundwater infiltration	+	+	+	+	+	+	+	+	+	+	+	+
Removal of pollutants from water	–	–	–	–	–	–	–	–	–	–	–	–
Direct food and fibre production	·	·	·	·	·	·	·	·	·	·	·	·
Keeping potential for agricultural land	–	–	–	–	–	–	–	–	–	–	–	–
Soil development and nutrient cycle	–	–	–	–	–	–	–	–	–	–	–	–
Preventing soil erosion	o	o	o	o	o	o	o	o	o	o	o	o
Cleaner air	–	–	–	–	–	–	–	–	–	–	–	–
Positive impact on land and property	–	–	–	–	–	–	–	–	–	–	–	–
Local distinctiveness	o	o	o	o	o	o	–	–	–	–	–	–
Sense of space and nature	o	o	o	o	o	o	–	–	–	–	–	–
Recreation	+	+	+	+	+	+	–	–	–	–	+	–
Opportunities for education, training and social interactions	+	+	+	+	+	+	+	+	–	–	+	+
Tourism opportunities	o	o	o	o	o	o	–	–	–	–	–	–

- functions that are to be developed, according to the provisions of the Warsaw Spatial Policy chapter referring to the WNS
- functions that are to be developed in selected areas of the WNS, according to the provisions of the Warsaw Spatial Policy chapters, referring to other than WNS issues (e.g. spatial structure, cultural heritage protection, water management)
- functions that are not mentioned, according to the provisions of the Warsaw Spatial Policy

An analysis of the Warsaw Spatial Policy also exposed some issues that should be raised from a methodological point of view. We argued that a land use approach, in which functions have been identified on the basis of the type of land use, seems to be insufficient for WGI (e.g. cultural functions, such as creating a landscape identity, could not be identified on the basis of a land use analysis). In order to analyse and plan functions of WGI elements, particular objects should be documented. Only for them it would be reasonable to analyse individually current functions, assess their significance, potential conflicts and finally plan for the future. Also an inventory of WGI functions should be redefined on the basis of its object documentation. For example, there might be different types of recreation, tourism or education possibilities depending on the object.

3.3. Connectivity principle

- Is the WNS planned as a coherent structure? What are the prospects of implementing the WNS as a coherent structure?

The WNS, according to its assumptions, was planned as a coherent structure. Its linkages with the regional ecological network (assured by the Vistula valley and the vast forest complexes) were also taken into account as an indispensable condition for the performance of the WNS (Figure 5). The integrity of the WNS was determined through the delineation of the core and supporting areas and the recommendations for the development of greenways (street plantations) (Figure 2). In several cases (especially in densely built-up areas), however, connectivity between individual areas of the WNS have still not been accomplished (the recommended linkages have not yet been completed). It should be stressed that over the course of the delineation of the WNS, a focus was placed on functional connectivity. The empirical evidence on whether and how the planned linkages function is still insufficient.

An analysis of planned spatial structure of the city as well as its transportation system allowed us to identify the presence of barriers (physical obstructions) and gaps (interruptions in continuity). So, in answering the question related to the prospects of implementation of the WNS as the coherent structure, we should raise the following concerns:

(1) The planned land use structure within the boundaries of the WNS is varied. In the WNS core areas, forests and parks prevail, although residential areas also exist. WNS supporting areas include nearly all urban functions. This implies that in some WNS supporting areas, land use other than the green and/or blue spaces are widespread (Figure 3); in such cases, connectivity may be achieved only through the development of different land cover features. Safeguarding and implementation of vegetation as the most welcome land cover in the case of structural and functional connectivity of the WNS may be addressed through planning measures by manipulating the size of the RBVA. As application of the RBVA is a quantitative rather than qualitative measure, physical connectivity should be developed by site-scale solutions. This issue is discussed in the *Multi-scale approach* section.

(2) Existing and planned main roads create physical barriers that affect WNS connectivity (Figure 4). The best WNS continuity conditions were found along the Vistula river corridor.

(3) The existence of the Vistula valley, forests and other open spaces in the vicinity of the city is enabling the development of connectivity at metropolitan and regional level (Figure 5).

- To what extent does the WGI follow the structure of the WNS? What sort of corridors and/or greenways could be/should be developed in order to enhance WGI connectivity?

To a certain extent the WGI follows the WNS structure. That is obvious because certain areas /land uses which form the WNS must also be considered as WGI objects. At this moment of analysis, it is too early to decide precisely which areas/objects will form the WGI structure at the city level, but main links, particularly within the core areas can be utilised for WGI. This ecological /environmental connectivity will solve half of the connectivity problem. The other half, according to GI connectivity rules, refers to movement of people between GI objects. In GI ideology, there is a strong belief that

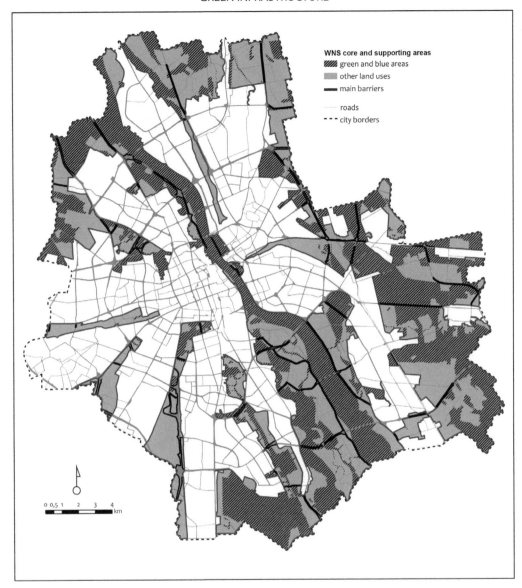

Figure 4. WNS connectivity problems. Source: Author.

this movement should be performed by foot or bicycle (sustainable transportation). Such systems are still under construction in Warsaw and at the moment, the priority is given to support transport not to connect existing parks and other green spaces.

3.4. Multi-scale approach principle

• Is the WNS concept based on the multi-scale principle?

The multi-scale approach to the implementation of the WNS formed one of the main assumptions on which this concept was based. According to the Warsaw Spatial Policy, the WNS is considered to be an element of the wider regional ecological network. WNS delineated at the city level was supposed to be implemented through spatial development of particular areas (land uses), according to certain rules presented in the Warsaw Spatial Policy. So, local plans since the beginning have been considered as the main measure of WNS implementation. Thus, to verify the adopted assumption, we analysed

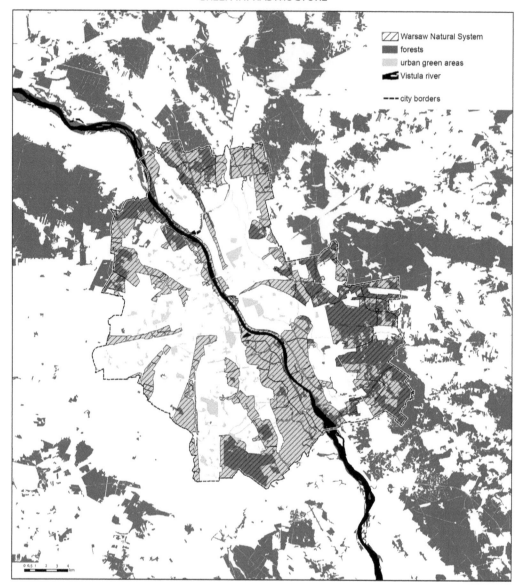

Figure 5. Connectivity of WNS with ecological network at the metropolitan level. Source: Author.

the distribution and spatial scope of local plans in force that are related to WNS issues. Information on local plans was acquired from the City of Warsaw's on-line map service.

The analysis revealed that local plans have been prepared only for certain areas of the WNS (Figure 6). Thus, only for these areas are recommended RBVA sizes maintained. Of course, complying with this recommendation does not guarantee a favourable result. This instead depends on green space layout and vegetation characteristics. Areas of the WNS for which no local plan has been developed may thus not be addressed. That creates one of the main problems of the WNS implementation.

- To what extent could/should WGI follow the WNS implementation rules?

Local plans, which are in fact the only instrument for WNS implementation, do not seem to be sufficient for WGI. Also taking into account multi-object principles (see Section 2.3.3) and depending on the

Figure 6. Local plans as a measure for WNS implementation. Source: Author.

respective scale, different sets of instruments for the implementation of WGI should and could be designed. Instruments for the implementation of WGI at the site scale could be considered bottom-up initiatives which have just appeared in Warsaw's different districts and neighbourhoods (e.g. the enhancement of green spaces in the framework of participatory budgets). On the other hand, at the city level a new programme called 'One million trees', similar to that carried out in New York, Chicago and other cities has been launched. All such initiatives need co-ordination and the WGI umbrella seems to be the right one.

3.5. Multi-object principle

The multi-object approach did not underpin the WNS concept. However, taking into account the types of land uses which can be developed within the WNS borders (see Figure 3) we identified those types which are in tune with the GI typology formulated by Davies et al. (2006).

The analysis revealed that while the WNS is diverse, not all planned land uses serve as potential GI elements at the city scale. Nevertheless, it should be stressed that GI is a hierarchal concept (see Section 4.4), and thus even in zones designated for land uses that are not part of city-scale GI, actual GI elements can be delivered at the site scale (e.g. green roofs, green walls).

A comparison between potential elements of WGI (based on the typology developed by Davies et al., 2006) with planned WNS land use categories reveals a number of gaps or divergences. GI land uses that are not planned for in the WNS include: agriculture areas (including all arable land for horticulture, orchards, allotments, community gardens and urban farms).

4. Conclusions: prospects for WGI

In this paper, we have discussed a possibility of GI idea implementation, taking as a point of departure the WNS concept, already existing in Warsaw's spatial planning tradition. Before presentation of prospects for WGI implementation, two important remarks should be made.

(1) Since the last decade of the twentieth century, the idea of the WNS and its precursors has been present in Warsaw's planning documents. Planners, public officers, decision makers accepted it and got used to it. That is why we suggest to preserve the WNS as a 'natural layer' on which WGI will be developed

(2) In our paper, in discussing possible WGI development, we have followed the principles presented by Hansen and Pauleit (2014), but in an order that is a bit different. This turned out to be more reasonable from the point of view of transforming the WNS into WGI.

4.1. Objects

WNS consists of different areas representing various land uses while WGI should be presented as a system of defined objects—protected, managed, redeveloped, all according to GI principles. The city must decide which of the objects are of strategic importance. Those should be a stable element of city spatial structure and will form WGI at the city level. It should also be decided what type of objects should be promoted at district and neighbourhood levels. It is obvious that for Warsaw, Vistula Valley and the Warsaw Escarpment will form the main elements of the WGI. But in order to manage those areas efficiently, precise borders of those elements should be established and agreed. There is also a decision to be made, regarding which parks, from the existing 88 ones, are of strategic importance. WGI plans are challenged by visions of spatial development that do not provide space for agriculture or allotment gardens. This issue needs to be discussed with Warsaw authorities, especially given the growing popularity of urban agriculture.

Residential areas and ways in which accompanying green spaces are designed, managed and maintained appear essential to WGI. Those areas directly influence the quality of life and can form connecting links between strategic WGI objects.

4.2. Multi-scale

WGI plans and strategies must adopt a multi-scale approach. There are two scenarios for possible concept implementation:

(a) The implementation of WGI is to be led by the municipal authorities based on a common conceptualisation of the citywide WGI. Amongst the relevant administrative bodies are the Environmental

Protection Department (formally responsible for the implementation of the WNS and green space policy) and the Architecture and Spatial Planning Department (formally responsible for the Warsaw Spatial Policy). In our opinion, the most important issue lies in the detailed analyses of the potential GI implementation measures, which are based on the spatial, social, economic and organisational conditions in Warsaw;

(b) The implementation of WGI is to be realised on the basis of local initiatives supported by various funding streams (e.g. community gardens, playgrounds and pocket parks). In such a case, the municipal authorities serve as an umbrella organisation that supports, advises and coordinates the local initiatives.

Of course, the predicted scenarios must not be viewed as competing with one another, but rather as complementary. Such approaches may thus be implemented simultaneously. However, this will depend on the political will of officials and on management practices in force in the City of Warsaw.

4.3. Connectivity

The implementation of this principle in the case of the WNS should be considered partially successful. The transportation network in Warsaw creates unavoidable barriers and gaps in a coherent WNS network. On the other hand, roads with accompanying street greenery, such as alleys, can be seen as linkages. Warsaw, however, is fortunate because a partly wild river valley crosses the city. Furthermore, the Warsaw Escarpment facilitates connectivity between the WNS elements. Special attention should be paid to different forms of planned connection (greenways and watercourses). While these connections are central to the WNS (and future WGI) functioning, the Warsaw Spatial Policy is too vague for these features to be established. Local plans (site level) seem to be crucial for fulfilling the connectivity principle.

4.4. Multifunctionality

For the WNS, it was limited to specific environmental functions. However, some social functions were assigned for selected areas included in the WNS. Key problems for WGI implementation lies in the evaluation of possible WGI objects from the point of view of their present and required functions: environmental and social, but in a more refined way.

4.5. Integration

This principle has been understood as an integration of different policy sectors relevant for GI implementation. This kind of integration has not been fully achieved in the case of the WNS due to its 'environmental' character and scope of the Warsaw Spatial Policy. Nevertheless, conducted analysis allowed us to consider not only the importance of policy sector integration (usually policy provisions refer to general rules and standards) but integration of these rules and standards on the particular GI area/object. It seems that integration should be performed not only at the level of policy sector but also at the level of particular objects, in their management rules and/or plans.

4.6. To sum up

On the basis of our analysis we can tell that Warsaw is able to face the challenge of GI implementation and the WNS may serve as a good point of departure for identification and characterisation of WGI objects. We also ponder the meaning and usefulness of 'old' concepts that have evolved from the planning tradition of different cities, as they are still significant and should not be put away. At the same time, we addressed and tested theoretical assumptions and principles of GI implementation. We proved their relevance but also found the need for their specifying and tuning.

Disclosure statement

No potential conflict of interest was reported by the authors.

Funding

This work was supported by the Ministry of Science and Higher Education of Poland [statutory research support].

References

Ahern, J. (1995). Greenways as a planning strategy. *Landscape and Urban Planning, 33*, 131–155. doi:10.1016/0169-2046(95)02039-V

Ahern, J. (2007). Green infrastructure for cities: The spatial dimension. In V. Novotny & P. Brown (Eds.), *Cities of the future: Towards integrated sustainable water and landscape management* (pp. 267–283). London: IWA Publishing.

American Rivers, Association of State and Interstate Water Pollution Control Administrators, National Association of Clean Water Agencies, Natural Resources Defense Council, The Low Impact Development Center, & U.S. Environmental Protection Agency. (2008). *Managing wet weather with green infrastructure action strategy.* U.S. Environmental Protection Agency. Retrieved from http://water.epa.gov/infrastructure/greeninfrastructure/upload/gi_action_strategy.pdf

Arendt, R. (2004). Linked landscapes creating greenway corridors through conservation subdivision design strategies in the northeastern and central United States. *Landscape and Urban Planning, 68*, 241–269. doi:10.1016/S0169-2046(03)00157-9

Barcelona Green Infrastructure and Biodiversity Plan. (2013). *Commission of urban habitat.* Barcelona: City Hall of Barcelona.

Benedict, M. A., & McMahon, E. T. (2006). *Green infrastructure—linking landscapes and communities.* Washington, DC: Island press.

Bryant, M. M. (2006). Urban landscape conservation and the role of ecological greenways at local and metropolitan scales. *Landscape and Urban Planning, 76*, 23–44. doi:10.1016/j.landurbplan.2004.09.029

Carter, T. (2009). Developing conservation subdivisions: Ecological constrains, regulatory barriers, and market incentives. *Landscape and Urban Planning, 92*, 117–124. doi:10.1016/j.landurbplan.2009.03.004

Chang, Q., Li, X., Huang, X., & Wu, J. (2012). A GIS-based green infrastructure planning for sustainable urban land use and spatial development. *Procedia Environmental Sciences, 12*, 491–498. doi:10.1016/j.proenv.2012.01.308

City of Chicago Green Stormwater Infrastructure Strategy. (2014). Retrieved from www.cityofchicago.org

Colding, J. (2007). 'Ecological land-use complementation' for building resilience in urban ecosystems. *Landscape and Urban Planning, 81*, 46–55. doi:10.1016/j.landurbplan.2009.03.004

Davies, C., MacFarlane, R., McGloin, C., & Roe M. (2006) *Green infrastructure planning guide. Version 1.1.* Green Infrastructure North West. Retrieved from http://www.greeninfrastructurenw.co.uk/resources/North_East_Green_Infrastructure_Planning_Guide.pdf

Dietz, M. E. (2007). Low impact development practices: A review of current research and recommendations for future directions. *Water Air Soil Pollution, 186*, 351–363. doi:10.1007/s11270-007-9484-z

European Commission. (2013). *Communication from the Commission to the European Parliament, the Council, the European Economic and Social Committee and the Committee of the Region. Green Infrastructure (GI)—enhancing Europe's natural capital (COM/2013/0249 final).* Retrieved from http://eur-lex.europa.eu/legal-content/EN/TXT/?uri=CELEX:52013DC0249

European Environment Agency. (2014). *Spatial analysis of green infrastructure in Europe* (Technical Report No 2/2014). EEA. Retrieved from http://www.eea.europa.eu/publications/spatial-analysis-of-green-infrastructure

Fleury, A. M. & Brown, R. D. (1997). A framework for the design of wildlife conservation corridors With specific application to southwestern Ontario. *Landscape and Urban Planning, 37*, 163–186. doi:10.1016/S0169-2046(97)80002-3.

Freeman, R. C., & Bell, K. P. (2011). Conservation versus cluster subdivisions and implications for habitat connectivity. *Landscape and Urban Planning, 101*, 30–42. doi:10.1016/j.landurbplan.2010.12.019

Giedych, R., Szulczewska, B., Dobson, S., Halounova, L., Doygun, H. (2014). Green infrastructure as a tool of urban areas sustainable development. In R. Davson, A. Wyckmans, O. Heidrich, J. Köhler, S. Dobson, & E. Feliu (reds.) *Understanding Cities: Advances in Integrated Assessment of Urban Sustainability* (pp. 94–108). Newcastle: Centre for Earth System Engineering Research.

Goddard, M. A., Dougill, A. J., & Benton, T. G. (2010). Scaling up from gardens: Biodiversity conservation in urban environments. *Trends in Ecology & Evolution, 25*, 90–98. doi:10.1016/j.tree.2009.07.016

Green Infrastructure Center and E² Inc. (2010). *Richmond green infrastructure assessment.* Green Infrastructure Centre. Retrieved from http://www.gicinc.org/PDFs/RichmondGIA_Report_FINAL.pdf

Green Infrastructure and Open Environments: The All London Green Grid. (2012). *Greater London authority.* London: Greater London Authority City Hall. ISBN 978-1-84781-505-7

Green Infrastructure Plan, City of Lancaster. (2011). Retrieved from http://cityoflancasterpa.com/sites/default/files/documents/cityoflancaster_giplan_fullreport_april2011_final_0.pdf

Green Infrastructure and Territorial Cohesion. (2011). *The concept of green infrastructure and its integration into policies using monitoring systems.* Luxembourg: European Environment Agency.

Hansen, R., & Pauleit, S. (2014). From multifunctionality to multiple ecosystem services? A Conceptual framework for multifunctionality in green infrastructure planning for urban areas. *Ambio, 43*, 516–529. doi:10.1007/s13280-014-0510-2

Hostetler, M., Allen, W., & Meurk, C. (2011). Conserving urban biodiversity? Creating green infrastructure is only the first step. *Landscape and Urban Planning, 100*, 369–371. doi:10.1016/j.landurbplan.2011.01.011

Ignatieva, M., Stewart, G. H., & Meurk, C. (2011). Planning and design of ecological networks in urban areas. *Landscape and Ecological Engineering, 7*, 17–25. doi:10.1007/s11355-010-0143-y

Jones, M., Howard, P., Olwig, K. R., Primdahl, J., & Herlin, I. S. (2007). Multiple interfaces of European landscape convention. *Norwegian Journal of Geography, 61*, 207–215. doi:10.1080/00291950701709176

Jongman, R. H. G., Külvik, M., & Kristiansen, I. (2004). European ecological networks and greenways. *Landscape and Urban Planning, 68*, 305–319. doi:10.1016/S0169-2046(03)00163-4

Kambites, C., & Owen, S. (2006). Renewed prospects for green infrastructure planning in the UK. *Planning, Practice & Research, 21*, 483–496. doi:10.1080/02697450601173413.

Land Use Consultants. (2009). *South east green infrastructure framework. From policy into practice.* The Southeast Green Infrastructure Partnership. Retrieved from http://segip.org/wp-content/uploads/2010/01/SEGIFramework.finaljul09.pdf

Landscape Institute. (2011). *Local green infrastructure. Helping communities make the most of their landscape.* Landscape Institute. Retrieved from http://www.landscapeinstitute.org/PDF/Contribute/LocalGreenInfrastructurewebversion_000.pdf

Madureira, H., Andresen, T., & Monteiro, A. (2011). Green structure and planning evolution in Porto. *Urban Forestry and Urban Greening, 10*, 141–149. doi:10.1016/j.ufug.2010.12.004

Maryland's Green Infrastructure Assessment and GreenPrint Program. (2004). Green infrastructure – linking lands for nature and people. Case study series. Retrieved from http://www.conservationfund.org/images/programs/files/Marylands_Green_Infrastructure_Assessment_and_Greenprint_Program.pdf

Mell, I. C. (2014). Aligning fragmented planning structures through a green infrastructure approach to urban development in the UK and USA. *Urban Forestry & Urban Greening, 13*, 612–620. doi:10.1016/j.ufug.2014.07.007

Mell, I. C., Henneberry, J., Hehl-Lange S., & Keskin B. (2013). Promoting urban greening: Valuing the development of green infrastructure investments in the urban core of Manchester, UK. *Urban Forestry & Urban Greening, 12*, 296–306. doi: 10.1016/j.ufug.2013.04.006

Natural Economy Northwest. (2009). *A Guide to planning green infrastructure at the sub-regional level DRAFT (v3.1).* Green Infrastructure North West. Retrieved from http://www.greeninfrastructurenw.co.uk/resources/Sub_Regional_GI_planning_guide_v3.pdf

Natural England. (2008). *The essential role of green infrastructure: Eco-towns green infrastructure worksheet. Advice to promoters and planners.* Town and Country Planning Association. Retrieved from http://www.tcpa.org.uk/data/files/etws_green_infrastructure.pdf

Niemelä, J., Saarela, S.-R., Söderman, T., Kopperoinen, L., Yli-Pelkonen, V., Väre, S., & Kotze, D. J. (2010). Using the ecosystem services approach for better planning and conservation of urban green spaces: A Finland case study. *Biodiversity Conservation, 19*, 3225–3243. doi:10.1007/s10531-010-9888-8

Novotny, V., Ahern, J., & Brown, P. (2010). *Water centric sustainable communities: Planning, retrofitting and building the next urban environment.* Hoboken, NJ: John Wiley & Sons.

NYC Green Infrastructure Plan, 2014 Annual Report. (2014). Retrieved from: http://www.nyc.gov/html/dep/pdf/green_infrastructure/gi_annual_report_2015.pdf

NYC Green Infrastructure Plan, A Sustaianle Strategy for Clean Waterways. (2010). Retrieved from: http://www.nyc.gov/html/dep/pdf/green_infrastructure/NYCGreenInfrastructurePlan_LowRes.pdf

Opdam, P., Steingröver, E., & Rooij, S. (2006). Ecological networks: A spatial concept for multi-actor planning of sustainable landscapes. *Landscape and Urban Planning, 75*, 322–332. doi:10.1016/j.landurbplan.2005.02.015

Osborn, F. V. & Parker, G. E. (2003). Linking two elephant refuges with a corridor in the communal lands of Zimbabwe. *African Journal of Ecology, 41*, 68–74.

Palomares, F. (2001). Vegetation structure and prey abundance requirements of the Iberian lynx: implications for the design of reserves and corridors. *Journal of Applied Ecology, 38*, 9–18. doi:10.1046/j.1365-2664.2001.00565.x

Parker, K., Head, L., Chisholm, L. A., & Feneley, N. (2008). A conceptual model of ecological connectivity in the Shellharbour Local Government Area, New South Wales, Australia. *Landscape and Urban Planning, 86*, 47–59. doi:10.1016/j.landurbplan.2007.12.007

Pauleit, S., Liu, L., Ahern, J., & Kazimierczak, A. (2011). Multifunctional green infrastructure planning to promote ecological services in the city. In J. Niemela (Ed.), *Urban ecology, Patterns, processes and applications* (pp. 272–285). Oxford: Oxford University Press.

Pyke, C., Warren, M. P., Johnson, T., LaGro, J. Jr., Scharfenberg, J., Groth, P., ... Main, E. (2011). Assessment of low impact development for managing stormwater with changing precipitation due to climate change. *Landscape and Urban Planning, 103*, 166–173. doi:10.1016/j.landurbplan.2011.07.006

Rutherford, S. (2007). *The green infrastructure guide. Issues, implementation strategies and success stories.* West Cost Environmental Law Research Foundation. Retrieved from http://wcel.org/sites/default/files/publications/The%20Green%20Infrastructure%20Guide%20-%20Issues,%20Implementation%20Strategies,%20and%20Success%20Stories.pdf

GREEN INFRASTRUCTURE

Spatial Planning and Development Act. (2003). *Ustawa o planowaniu i zagospodarowaniu przestrzennym* [Law on spatial planning and development]. Official Journal of Law No. 80/717.

Statistical Office in Warsaw. (2015). *Panorama Dzielnic Warszawy w 2014* [Panorama of Warsaw Districts in 2014]. Statistical Office in Warsaw. Retrieved from http://warszawa.stat.gov.pl/publikacje-i-foldery/inne-opracowania/panorama-dzielnic-warszawy-w-2014-r-,5,16.html

Szulczewska, B., & Kaftan, J. (Eds.). (1996). *Kształtowanie Systemu Przyrodniczego Miasta [Shaping of urban natural system]*. Warsaw: IGPiK.

Szulczewska, B., Giedych, R., Borowski, J., Kuchcik, M., Sikorski, P., Mazurkiewicz, A., & Stańczyk, T. (2014). How much green is needed for a vital neighbourhood? In search for empirical evidence. *Land Use Policy, 38*, 330–345. doi:10.1016/j.landusepol.2013.11.006

The Interior Green Belt: Towards an urban green infrastructure in Vitoria-Gasteiz. Working document. (2012). Environmental Studies Centre. Retrieved from www.vitoria-gasteiz.org/ceac

The Royal Institution of Chartered Surveyors. (2011). *Green infrastructure in urban areas. RICS practice standards*. Retrieved from RICKS: http://www.joinricsineurope.eu/uploads/files/GreenInfrastructureInformationPaper1stEdition.pdf

Viles, R. L. & Rosier, D. J. (2001). How to use roads in the creation of greenways: case studies in three New Zealand landscapes. *Landscape and Urban Planning, 55*, 15–27. doi:10.1016/S0169-2046(00)00144-4

Warsaw Environmental Study. (2006). *Opracowanie Ekofizjograficzne*. Warsaw Architecture and Spatial Planning Department. Warsaw Municipality. Retrieved from http://architektura.um.warszawa.pl/ekofizjografia

Warsaw Spatial Policy. (2006, amended 2010, 2014). *Studium Uwarunkowan i Kierunkow Zagspodarowania Przestrzennego m.st. Warszawy, aktualizacja*. Warsaw Architecture and Spatial Planning Department. Warsaw Municipality. Retrieved from http://architektura.um.warszawa.pl/studium

Siting green stormwater infrastructure in a neighbourhood to maximise secondary benefits: lessons learned from a pilot project

Danielle Dagenais ⓘ, Isabelle Thomas and Sylvain Paquette ⓘ

ABSTRACT

When siting green stormwater infrastructure (GSI), cities do not respond only to technical and regulatory requirements; they also strive to maximise environmental, aesthetic and social benefits. To help cities optimise the siting of GSI in the context of climate change, we developed a participatory decision support tool. Applied to a neighbourhood, this tool identified only a few sites where GSI would yield all secondary benefits and reduce climate change vulnerability. In the light of the need for large-scale implementation of GSI in cities, this finding raises the following questions: How can the potential benefits provided by a site be best identified? Are there potential synergies or antagonisms between benefits? How do they relate to vulnerability? Can a participatory decision-making process involving local stakeholders improve this process? Informed by the existing literature on balancing ecosystem services and vulnerability, these questions are addressed within a broader perspective of landscape design and urban planning.

1. Introduction

In the coming years, the conjunction of climate change and urbanisation will exacerbate current stormwater management problems in cities. Expected increases in volume and peak flow will result in higher levels of backflow, overflow and erosion, as well as the deterioration of the water quality of receiving watercourses (Astaraie-Imani, Kapelan, Fu, & Butler, 2012). Because of this, green stormwater infrastructure (GSI) is being used to add to or even replace original grey infrastructure (Ellis, 2012; Fletcher et al., 2015; Keeley et al., 2013). Unlike grey infrastructure, GSI plays a role not only in the quantity of stormwater (volumes and peak flows), but its quality as well (Ellis, 2012; Hunt, Davis, & Traver, 2012). GSI is defined as

> an approach that communities can choose to maintain healthy waters, provide multiple environmental benefits and support sustainable communities. Unlike single-purpose grey stormwater infrastructure, which uses pipes to dispose of rainwater, green infrastructure uses vegetation and soil to manage rainwater where it falls. (US EPA, 2012 in Fletcher et al., 2015, p. 532)

Examples of GSI in current use are green roofs, bioretention areas, swales, and vegetated detention basins. Interest in GSI is high, particularly in North America (USA and Canada), Northern Europe, and Oceania (Australia and New Zealand) (Dagenais, Paquette, Fuamba, & Thomas-Maret, 2011; Fletcher et al., 2015; Keeley et al., 2013). In Canada in particular, GSI has been employed in several cities

(Marsalek & Schreier, 2009). GSI can be very appealing to cities because it can be easily integrated in new developments or retrofitted into the urban fabric where space is limited and valuable (Rodenburg & Nijkamp, 2004). It also provides a substantial number of benefits (Novotny, Ahern, & Brown, 2010).

1.1. Benefits of GSI and its impact on vulnerability

GSI's primary benefit of hydrological performance and improved water quality is supplemented by secondary benefits that play a role in climate change adaptation and in the quality of urban spaces. According to the literature, GSI can alleviate heat islands, reduce biodiversity losses, support physical and mental health, and improve the population's general quality of life (Gill, Handley, Ennos, & Pauleit, 2007; Keeley et al., 2013). These benefits are also referred to as ecosystem services or functions (Everard & McInnes, 2013; Moore & Hunt, 2012; Novotny et al., 2010). The secondary benefits provided by GSI also contribute to its enhanced social acceptability and the interest taken by stakeholders such as municipalities, professionals, developers, and the broader public in GSI implementation (Dagenais et al., 2011; Fryd, Jensen, Ingvertsen, Jeppesen, & Magid, 2010; White & Alarcon, 2009). Furthermore, cities emphasise these secondary benefits or functions in their rationale for implementing GSI (Dagenais et al., 2011). Moreover, due to the multifunctionality of GSI, its duplication of grey infrastructure functions, and its reduction of heat islands, several authors also highlight the fact that GSI helps reduce cities' vulnerability and thereby increase their resilience to climate change (Ahern, 2011; Pyke, Warren, Johnson, LaGro, & Scharfenberg, 2011; Novotny et al., 2010). Note that the vulnerability results from exposure to risk, sensitivity to this risk, and the absence of the capacity to adapt (Dagenais et al., 2012; Maret & Cadoul, 2008).

The concepts of secondary benefits, multifunctionality, and ecosystem services, as well as those of vulnerability and resilience, arise from different types of concerns and are derived from distinct frames of references (e.g., improvement of the living environment, sustainable development, risk management). According to some authors, the field of urban planning now needs to merge these frameworks together and take all of these concerns into account simultaneously in order to design and create more sustainable and resilient cities (Ahern, 2011; Hansen & Pauleit, 2014). This goal is applicable to the implementation of GSI (Novotny et al., 2010).

1.2. Implementing GSI to increase secondary benefits and reduce vulnerability

To take secondary benefits and vulnerability into account when planning and implementing GSI, sites that are amenable to furnishing secondary benefits and reducing vulnerability to climate change need to be located. However, cities are not well equipped with respect to the tools and the expertise required for locating these sites.

Several decision aid tools have been developed to support the implementation of GSI within a territory (Dagenais et al., 2013). Some of these are meant for choosing the type of infrastructure that will be adequate for a predetermined site, rather than attempting to determine the right location within an urban territory for the installation of adapted GSI (e.g., Rivard, 2011; Toronto & Region Conservation Authority, Credit Valley Conservation Authority [TRCA/CVCA], 2010). However, it is important for cities to choose the location of this type of infrastructure carefully, not only to ensure optimal hydrological performance and maximise water quality, but also to take advantage of the expected secondary benefits (Everard & McInnes, 2013; Fryd et al., 2010; Keeley et al., 2013; Uzomah, Scholz, & Almuktar, 2014). Reducing the vulnerability and increasing the resilience of cities to climate change is another goal that should also be considered (Ahern, 2011). However, the existing tools do not take all of these objectives into consideration at once.

Finally, both reducing vulnerability and considering secondary benefits require integrating different kinds of knowledge and thought processes so that all of the potential outcomes and implications of implementing GSI in a given territory can be understood. Information sharing and collaborative knowledge building between professionals of various disciplines and researchers, as well as between specialists, elected officials, and the general public, is necessary for successful implementation (Després, Vachon, & Fortin, 2011; White & Alarcon, 2009). Among the existing methods, a participatory process is often chosen to ensure that various knowledge sources are integrated into a project, such as in the case of the development of a decision support tool adapted to the needs of end users like city planners, engineers, and landscape architects (Després et al., 2011; Robertson & Simonsen, 2012).

In summary, this article addresses the following two questions. First, how can GSI installation sites capable of providing both secondary benefits and reducing climate change vulnerability at the neighbourhood level be identified? Second, how can considerations associated with secondary benefits and climate change vulnerability be incorporated into a decision support tool for the implementation of GSI at this scale? The article describes the development of such a decision support tool that (1) maps and takes secondary benefits into account, including social benefits, (2) integrates vulnerability reduction into the implementation of GSI, and (3) has an iterative and participative elaboration process. The article also highlights the observations and questions arising from the pilot application of the tool within an iterative, participative multidisciplinary workshop on the subject of the Beauport borough of the City of Québec in the Province of Québec, Canada. The decision tool is destined to be used by a variety of professionals potentially involved in the implementation of GSI and, eventually, by the general public.

2. A complex undertaking

The decision support tool was developed as part of a multidisciplinary research project that took place between 1 March 2010 and 31 March 2012. The project was particularly aimed at making the most of the secondary benefits and the reduction of vulnerability to climate change associated with GSI within a decision support tool that would be useful to planning professionals involved in its implementation.

The project consisted of 6 components, each of which played a part in the creation of the decision tool (Figure 1):

(1) An international case study of four cities or neighbourhoods that had successfully implemented GSI on a large scale: Portland, USA; Kronsberg, Germany; Philadelphia, USA; and Toronto, Canada to identify secondary benefits emphasised by cities in their rationale for implementing GSI and the problems that had an effect on the city's determination to implement them (Dagenais et al., 2011, 2012) (Table 1);

(2) The application of a new identification method for social and territorial sensitivity to be used at the local (neighbourhood, city) scale (Thomas et al., 2012);

(3) The determination and mapping of zones that could give rise to secondary benefits at the neighbourhood scale;

(4) Simulation of the maximal implementation of GSI within particular sectors of a neighbourhood (a residential street, a portion of an industrial sector, and a large commercial centre);

(5) The development of a prototype decision support tool;

(6) Testing of the tool in a pilot sector, the borough of Beauport, in the City of Quebec, Canada during a multidisciplinary workshop (Dagenais et al., 2012).

Components 2 to 5 were carried out using the literature and the terrain, which was a pilot sector of the borough of Beauport. Due to the importance of this sector to the development of the decision tool, it is described in Section 2.1 prior to the description of the tool itself.

2.1. The territory used for the pilot project

The decision tool was developed for the borough of Beauport, which is in an already built-up sector of the City of Québec in the Canadian province of Québec (Figure 2). It covers a territory of 7334 km^2 and its population was 77 905 in 2011 (City of Quebec, 2014). The site is 51% impervious, and has a separate drainage system (Dagenais et al., 2012). Beyond the willingness of its public officials to participate in

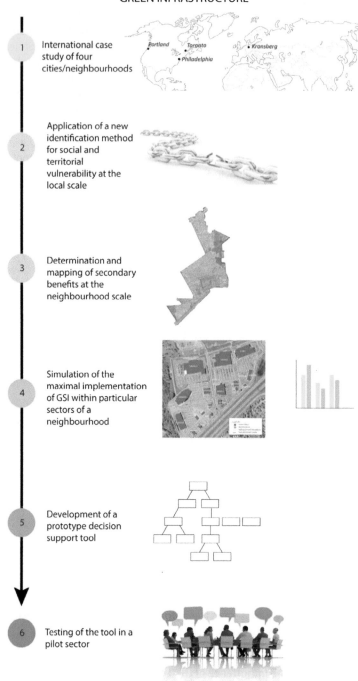

Figure 1. Outline of the project steps.
Source: A. Leboeuf.

this process, Beauport was chosen because it includes industrial, commercial and residential zones (Figure 2). Each of these zones differs in terms of water quality, available space, property ownership and

Table 1. Expected secondary benefits in cities.

Case	Portland	Kronsberg	Philadelphia	Toronto
SECONDARY BENEFITS				
Aesthetic	x	x	x	x
Air Quality	x			x
Biodiversity and habitats	x	x	x	x
Economic Activity			x	x
Education	x	x	x	x
Energy Savings	x	x	x	x
Greening		x	x	x
Identity	x			x
Quality of the Living Environment	x		x	x
Recreation			x	x
Safety/ Multimodal Transportation Enhancement	x	x	x	
Urban Heat Island Mitigation	x	x	x	x

Source: Dagenais et al. (2012), p. 73.

division. The borough level was chosen for the application of the tool because the borough is the entity responsible for installing decentralised rainwater management systems such as GSI on this territory.

2.2. The decision support tool

After an in-depth and exhaustive literature review of existing tools, we decided to develop a tool in the form of a decision tree supported by maps (Dagenais et al., 2012, 2013). Decision trees are easy to use and are employed in many different fields (such as, botany, engineering) (Gorunescu, 2011). Furthermore, decision trees are already used to select the appropriate GSI for a given site in some stormwater manuals (e.g., TRCA/CVCA, 2010). For their part, maps are part of the basic 'tool kit' of the professionals most likely to be involved in GSI implementation—planners, hydrologists, and ecologists (Ndubisi, 2002). Maps are easy to manipulate, and residents can understand them. When shared, they encourage dialogue between stakeholders (McCall & Dunn, 2012).

The decision tool consists of the four successive steps detailed below, which allow the installation sites to be targeted and the most appropriate type of GSI for the site to be selected. Each of the steps of the decision tool requires the creation of one or several maps using a geographic information system (GIS) (Figure 3). These superimposed maps make it possible to identify potential intervention sites. A diagram of the tool and the maps required is shown in Figure 3. The tool's steps are the following:

Step 1. Exclusion of sectors that cannot be considered for GSI implementation (Figure 3, step 1)

The reasons for exclusion are either that the sectors already contribute to stormwater management (natural drainage lines, riparian buffers, wetlands, woodlands) or that the use of GSI would not be appropriate according to the technical criteria found in relevant stormwater management manuals. In our case, the manuals that were relevant to Southern Québec were the *Low Impact Development (LID) Stormwater Management Manual* of the nearby Toronto Region (TRCA/CVCA, 2010) and the Stormwater Management Guide from the Ministry of Sustainable Development, the Environment, and Climate Change Efforts of the province of Québec by Rivard (2011). For example, areas where the slope was too steep (slope ≥ 5%) were excluded to ensure that the flow rate of the water entering the bioretention area was low enough to allow proper filtration and infiltration of the stormwater (Rivard, 2011), as well as areas where the water table was too high (≤1.2 m from the bottom of the GSI) and pollution hotspots (TRCA/CVCA, 2010). The exclusion of sectors is described in the LID planning practices, an approach to site design that aims for the at-source management of stormwater and the minimisation of transformations on the site during interventions so as to conserve the hydrological features present before development (Fletcher et al., 2015).

Step 2. Identification of sectors requiring a specific intervention for one or more problems: overflow, backflows, aquifer recharge (Figure 3). This step involves the precise identification of the origin of the problems to be resolved. When the objective is to restore or maintain the water quality of a

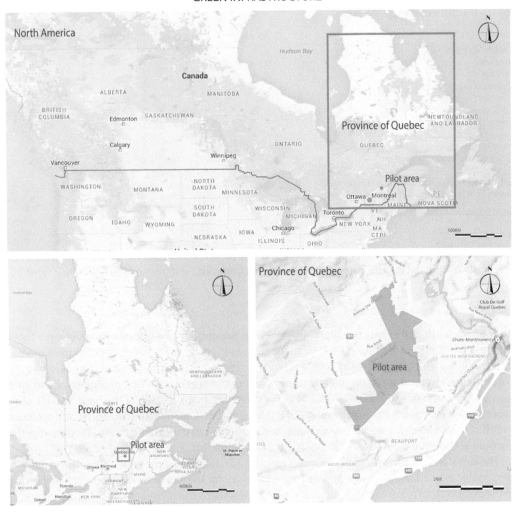

Map Key

Pilot area

Figure 2. Location of the pilot area, the borough of Beauport.
Source: Google Maps, adapted by Leboeuf and Besson.

watercourse, an entire drainage basin may be identified as the intervention sector (Dagenais et al., 2011; Walsh, Fletcher, & Ladson, 2005). Note that the problems have an effect on the city's determination to implement these systems as well as on the choice of GSI locations within the territory (Dagenais et al., 2011; TRCA/CVCA, 2010).

Step 3. Within these sectors, the priority areas for implementation are identified based on potential secondary benefits, hazards, and territorial and social sensitivity (vulnerability). The sectors that are most likely to provide secondary benefits are mapped differently depending on the type of benefit: environmental, or social and aesthetic. The sectors with the potential to furnish environmental benefits can be mapped using data on environmental issues. For example, potential heat islands were identified using maps created by the Province of Québec's Public Health Institute showing areas with an elevated daytime surface temperature (INSPQ & CERFO, 2012; Figure 3, step 3).

Step of the Decision Tool	List of Maps	Example Map
1. Sectors to exclude • Sites that might generate heavily polluted runoff due to the production, storage or handling of fuel or other hazardous material • Shallow water table (≤ 1.2 m from the bottom of the GSI) • Steep slopes (≥ 5%) • Natural areas of interest, woods • Stream buffer strips (e.g. 20 m from the high water mark for the City of Québec)	• Map of natural areas of interest, woods • Map of water table depths • Map of current and past land use, identifying sites that could generate heavily contaminated runoff • Map identifiying slopes steeper than 5% • Map of hydrologic soil groups	Excluded sectors Source:Google maps, adjusted by A Spector, L Besson
2. Sectors requiring a specific intervention to solve a problem • Overflow • Backflows • Aquifer recharge • Erosion • Water quality	• Map of aquifer area requiring recharge • Map of problem areas with respect to the targeted objective. • Map of sub-basins in major and minor drainage systems, identifiying the sub-basins responsible for problems	Overflows are not shown on the map for legal reasons Source Google maps, adjusted by A Spector, A Leboeuf
3. Priority areas based on potential secondary benefits, hazards, and territorial and social sensitivity • Vulnerability • Hazards; e.g.heat island effect • Territorial sensitivity • Social sensitivity	• Map of territorial and social sensibility • Map of heat islands • Map of highly impervious areas • Map of sector at risk for traffic accidents, especially involving pedestrians • Map of travel corridors to be promoted • Map of urban planning goals and location of sectors targeted for requalification • Map of infrastructure work planned for public land that could be combined with GSI implementation • Map of public land	Heat island effect map Source: INSPQ and CERFO (2012)
Composite Map		
4 Criteria for choosing GSI location and type • Source of runoff (runoff quality) • Availability of space to implement GSI • Impervious or pervious surface • Presence of suitable roofs for green roofs (slope < 10%) • Possibility of redirecting runoff to GSI farther from source	• Aerial map to assess the presence of flat roofs, available space, type of surface, etc. • Map of hydrologic soil groups	Aerial map Source: Google maps

Figure 3. Schematic representation of the decision tool with list of the maps created. In black, maps created for the workshop. In grey, additional maps suggested by the participants.

For aesthetic and social types of secondary benefits (Table 1), the city's planning objectives and various additional documents can be used. This makes it possible to identify the territorial issues and the urban areas most likely to serve as GSI installation sites and provide the type of secondary urban

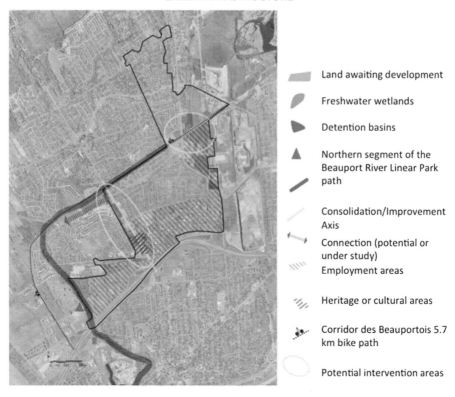

Land awaiting development

Freshwater wetlands

Detention basins

Northern segment of the
Beauport River Linear Park
path

Consolidation/Improvement
Axis

Connection (potential or
under study)

Employment areas

Heritage or cultural areas

Corridor des Beauportois 5.7
km bike path

Potential intervention areas

Figure 4. Map of aesthetic and social secondary benefits using urban planning objectives for the pilot area.
Source: City of Québec planning documents and Google Maps, adapted by L. Besson.

functions that these systems can ideally fulfil (e.g., cultural and heritage districts, travel corridors like linear parks, urban security, air quality, and education). These documents convey a sense of the principal public concerns and aspirations that have gained a level of consensus within the target community. This approach draws on the work of Fryd et al. (2010) in a SUDS implementation project in Odense, Denmark where hydrological, environmental, and sociocultural aspects were integrated. In that project, the sociocultural aspect was documented by analysing planning documents. In the case of the borough of Beauport, the results of a field visit held in May 2011 as well as different planning goals found in the Québec Urban Community Planning Diagram, the Planning and Development Master Plan, and various complementary documents were used (Dagenais et al., 2012) (Figure 4).

The focus of this third step is the application of a new ranking and mapping method for social and territorial sensitivity to climate change developed by Thomas et al. (2012). The main method involves four major substeps. The first one is to map out areas that are frequently subject to hazards (e.g., backflows, heat). Then, scenarios showing possible future hazards in a climate change context are established. The analysis of social sensitivity and territorial sensitivity serves as a starting point for the prioritisation of areas where appropriate adaptive tools like GSI should be put in place. Based on a method developed by Cutter, Boruff, and Shirley (2003), social sensitivity is measured through a number of indicators (e.g., percentage of the population aged 75 and over). In the context of hazards linked to climate change, elderly persons, young children, and recent immigrants, for instance, are considered to be vulnerable populations. The choice of the indicators and their rank is done based on the literature as well as the local context, which is determined through a multidisciplinary workshop with local experts. After an analysis of the principal components, the values obtained for each component are added up for each dissemination area and mapped (Figure 5(a)). The territorial sensitivity map was created using

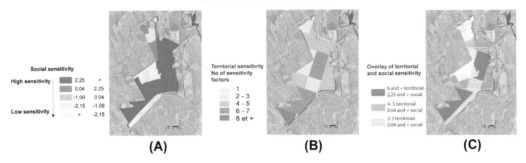

Figure 5. Social Sensitivity, Territorial Sensitivity, and Overall Sensitivity (a) Social sensitivity index map of the borough of Beauport by dissemination area, based on data from the 2006 Statistics Canada census. (b) Composite map of territorial sensitivity for the borough of Beauport created by aggregating components for each 400 m-wide parcel. (c) Global sensitivity map of the borough of Beauport. Source: Chanthamala, 2012 in Dagenais et al., 2012, Appendix A.3, p. 15, adapted by A. Leboeuf.

a method similar to that of D'Ercole and Metzger (2009). The aim is to identify major infrastructure at risk, like hospitals, schools, prisons, and major roads. Again the values obtained for each component are added up for each dissemination area and mapped (Figure 5(b)). Lastly, by superimposing the territorial and social sensitivity maps with maps showing hazards (Figure 5(c)), high-priority intervention areas can be identified.

Step 4. The type of infrastructure to be installed and precise installation locations for green infrastructure in priority areas are chosen in accordance with technical criteria (Figure 3, Step 4 and Table 2). The technical criteria were found in the relevant Stormwater Management Manuals mentioned above. The implementation of at-source GSI, such as green roofs and bioretention, was prioritised in the light of the space constraints associated with a built-up area. Drainage area, water quality, soil permeability, the presence of flat roofs, and the available permeable and impermeable spaces including the presence of buried infrastructure all determine which at-source GSI was used and where: green roofs, lined or unlined bioretention. This part of the decision tool was tested and perfected during spatial and hydrological simulations of maximal GSI implementation in the borough of Beauport.

2.3. The participative decision tool design workshop

The decision tool presented above was refined by means of a one-day multidisciplinary workshop that took place on 27 January 2012, organised with the cooperation of the City of Québec. The goal of this workshop was to test the decision tool under varied conditions, both to improve it and to promote information sharing and collaborative knowledge building between participants. The tool was submitted to a group of 20 professionals with a variety of backgrounds who were likely to be involved in decisions concerning GSI implementation within the borough of Beauport (engineers specialising in urban drainage, urban planners, a zoning specialist, a housing commissioner, environment specialists, and landscape architects).

The first part of the workshop consisted of a presentation on the background of the project and GSI and its benefits. During the second part, the professionals were divided into four multidisciplinary teams. The decision to use multidisciplinary teams was based on the results of the case studies and a literature review that demonstrated that multidisciplinary teams were key to the successful implementation of GSI (Dagenais et al., 2012, 2011; Fryd et al., 2010). Each team had to use the tool with a different hydrological objective in mind: reducing overflow, reducing backflows, promoting infiltration to recharge the aquifer, or controlling erosion or flooding in receiving streams. This allowed the decision tool to be tested with respect to different issues associated with stormwater management. Next, the teams had to use the decision-support tool to identify locations for GSI installations taking into consideration secondary benefits, vulnerability, and technical criteria as outlined in the decision support tool (Figure 3). At the end

Table 2. Technical criteria for the choice of GSI type and location in priority zones.

Green stormwater infrastructure	Type	Maximum drainage area	Runoff source	Size	Soil infiltration capacity	Maximum slope of the GSI	Runoff reduction estimate
Green roof	Extensive	N/A	Rain, roof	Roof area	N/A	Preferably flat roofs, otherwise with slope <10% (TRCA/CVCA, 2010)	45–55% (TRCA/CVCA, 2010)
	Intensive						
Bioretention	Complete infiltration	1 Ha	Roofs, lawns, green spaces, low traffic residential roads	5 to 10% of the drainage area, minimum 20 m^2	25 mm/h and up (water infiltrates within 48 h)	2:1 or less for the slopes inside the bioretention area itself in a low density development; vertical sides in high density development (TRCA/CVCA, 2010)	85% (no under drain (TRCA/CVCA, 2010)
	Filtration with partial recharge (underdrain)				Soil with limited infiltration capacity		45% (underdrain) (TRCA/CVCA, 2010)
	Filtration with partial recharge + elevated underdrain - For water with high nutriment concentrations or high volume				N/A		
	Filtration only		Parking lots, industrial areas not excluded previously, commercial and residential areas, medium to high traffic roads		Soil with very low infiltration capacity		

Source: Dagenais et al., 2012, Appendix A.2, p. 6 and Rivard, 2011, except where indicated.

Figure 6. Composite map. Superposition of heat islands, overall sensitivity, and urban planning objectives for the study area. Superposition of maps: Lucile Besson, adaptation by A. Leboeuf. Overflows are not shown on the map for legal reasons.

of the second part, each team had to make choices concerning implementation, and justify them. At the end of the workshop, the whole group was assembled to exchange their thoughts on the decision tool.

3. Results and discussion: lessons learned from the pilot project

The maps and decision tool presented above (Section 2.2, Figure 3) constitute the results of the project. Specifically, the testing of the decision tool prototype on a sector of the territory of the borough of Beauport led to the identification of two priority areas for implementation (Figure 6) where it was possible to provide all of the desired benefits while also targeting areas with the highest territorial and societal sensitivity. These sites can be considered as priority zones for the first GSI to be implemented within a territory. However, as mentioned in the Introduction, it will be necessary to implement GSI at a large number of sites in order to reach targets for volume reduction and peak flows at the neighbourhood scale. How do we identify more sites for the implementation of GSI? Are there synergies and antagonisms between benefits that we should take into account? And between benefits and vulnerability? How do we prioritise implementation at different sites?

Research on ecosystem services has identified land use as a common determinant for many services or environmental benefits. (Bennett, Peterson, & Gordon, 2009; Setälä et al., 2014). Changing land use can generate many benefits. For this reason, a second round of priority GSI implementation could involve implementing GSI in large impermeable areas like commercial centre parking lots. These areas generate great volumes of stormwater, exhibit high peak flow, and are heat islands (Figure 6) (Wilson, Hunt, Winston, & Smith, 2015). Their unsealing would also provide a range of environmental services such as air quality, biodiversity (e.g., pollination) and habitats, hydrological cycle regulation, microclimate benefits, carbon sequestration, and energy savings (Entrix, 2010; Gill et al., 2007; Setälä et al., 2014). Moreover, large impermeable surfaces are often the property of a single owner, which can simplify the negotiation for GSI implementation on these lands, in contrast to groupings of residential properties (Fryd et al., 2009). Taking advantage of correlations between benefits is a way to maximise the attainment of secondary benefits.

If more GSI is needed, our results indicated that areas of higher territorial and social sensitivity could be targeted. In the borough of Beauport, areas of high sensitivity included highly impermeable areas associated with high heat island effects, but extended beyond these areas (Figure 6). This is not surprising since research has shown that sensitive populations are often located in high surface temperature zones with little vegetation and high levels of imperviousness (Huang, Zhou, & Cadenasso, 2011). This is why many researchers advocate for an increase in vegetation cover in low-income neighbourhoods to increase the provision of ecosystem services (Clarke, Jenerette, & Davila, 2013). Similarly, infrastructure that determines territorial sensitivity, like hospitals, is also surrounded by impermeable zones. More research on the relationship between environmental benefits and vulnerability is needed. A hypothesis can be made that interventions aimed at reducing social and territorial vulnerability might allow such vulnerability to be addressed, along with increasing environmental benefits.

Our results show little spatial convergence between social, aesthetic and environmental benefits and the vulnerability of the territory studied (Figure 6). In this project, the social benefits were identified using planning documents. These documents summarise the population's aspirations and might target certain areas too precisely. The necessity of implementing GSI on more sites creates the requirement of, firstly, a more detailed analysis of the territory, and, secondly, trade-offs between options. The identification of sites capable of providing certain social benefits like recreation, safety, the enhancement of active transportation, and even education can be based on data compiled by municipalities. Other social benefits, like identity, quality of the living environment, and aesthetics, depend not only on the environment, but on the individuals and populations destined to live in contact with GSI. In that case, the identification of installation sites based on these benefits should be done in concert with this population (Low, Taplin, & Scheld, 2005). The local population and other stakeholders must also be invited to express their opinions on the prioritisation of the types of benefits to be provided. As illustrated by research on ecosystem services, different populations value different benefits (Martín-López et al., 2012). Various prioritisation methods exist, such as the formulation of participative scenarios that '(1) can create different visions of the future of the system, addressing its uncertainty and the main ecosystem services trade-offs, and (2) can propose consensual management strategies to determine a path toward a desirable future' (Palomo, Martin-Lopez, Lopez-Santiago, & Montes, 2011, p. 23). This approach allows for a qualitative and monetary evaluation of various planning scenarios and trade-offs that is easily understood by the stakeholders involved (Palomo et al., 2011).

As predicted, the use of the decision tool during a participatory workshop enabled knowledge sharing, particularly among the multidisciplinary team, and secondarily between all the participants and the researchers. It was clear that the participants were very interested in these discussions, which occur very rarely between professionals from different fields. The participants pointed out important issues and data related to planning for GSI installation that we, the researchers, had not identified, like the accessibility and exchange of data among departments, historical land use, and legal descriptions of systems located on private property.

Through their knowledge of the district (planning objectives, works planned) some participants could readily identify installation sites that, in their opinion, would result in multiple benefits for the community, without explicitly detailing these benefits (Dagenais et al., 2013). The professionals who participated also tended to take into account only the benefits that were either more familiar, of more concern to them, or that were frequently covered in the media, like the reduction of heat islands, to the detriment of benefits less closely related to their professional concerns, such as biodiversity, aesthetics, and quality of life. In fact, it would be useful to explain certain concepts more thoroughly, such as vulnerability (sensitivity) in particular. For this purpose, a number of training sessions should be scheduled before starting the participatory planning process so that all of the participants have a shared understanding of concepts and important background knowledge. The participants themselves also recommended this. Adding a wider range of participants (such as other professionals, experts, and residents) would increase the opportunities for issues neglected by the professionals directly involved in infrastructure installation to be taken into account. Finally, the categories of benefits included should correspond to categories that are meaningful for the participants.

4. Conclusion

The use of the decision tool in a pilot territory, the borough of Beauport on the outskirts of the City of Québec, Canada, allowed two priority GSI installation sites to be identified that are capable of providing the environmental, aesthetic and social benefits considered in this project and reducing the vulnerability of the most vulnerable zones. However, for hydrological and water quality results to be obtained on a city-wide scale, it will be necessary to install more GSI, some of which will be on sites where attaining all of these objectives will not be feasible. In such cases, it is important to understand the synergies and antagonisms between various benefits and between these benefits and vulnerability.

In fact, the relationships between the environmental benefits potentially associated with a given site are starting to be understood and formalised in the field of ecosystem services, especially the relationship between the common factor of land use and the potential environmental benefits (Bennett et al., 2009). Research has demonstrated a convergence of socially and territorially sensitive sectors and sectors exposed to environmental risks related to climate change (Huang et al., 2011). However, significant work still needs to be done so that the existing research results on the relationships between various social benefits, (e.g., the relationship between the aesthetic appreciation of landscape and the psychological and physical health of populations) or between social benefits, environmental benefits, and the reduction of vulnerability can be integrated into the implementation of GSI (Barthel, Folke, & Colding, 2010; Daniel et al., 2012; Staats, 2012). A more thorough understanding of the intensity of benefits and the influence of context also needs to be obtained (Pataki et al., 2011). Finally, the improvement of the design of the GSI itself in relation to a given site so that the performance expected as well as additional secondary benefits will be produced is an important goal (Hunt et al., 2012; Moore & Hunt, 2012).

Choices (trade-offs) between these benefits also need to be examined. Consultation with local populations and the development and presentation of scenarios clearly demonstrating the choices to be made between the benefits would promote dialogue and compromise between involved parties and produce more successful GSI implementation (Martín-López et al., 2012; Palomo et al., 2011). Finally, the capacity of designers and planners to overcome trade-offs and offer innovative solutions that can reconcile benefits or promote synergies between these benefits and reduce vulnerability can be put to full use (Ahern, Cilliers, & Niemelä, 2014).

Acknowledgements

The authors would like to express their gratitude to Ouranos for providing the funding for this project and for the production of this article, the City of Québec for organising the workshop, all the students who participated in the project, and all participants in the workshop for their invaluable contributions. The names of the latter are mentioned in the report quoted in this article (Dagenais et al., 2012). Finally, we wish to thank Prof. Tim D. Fletcher for his helpful comments on some passages of the article, and Andréanne Leboeuf for her help with the graphics and maps.

Disclosure statement

No potential conflict of interest was reported by the authors.

Funding

This work was supported by Ouranos, Consortium on Regional Climatology and Adaptation to Climate Change, Montreal, Quebec, Canada [grant number 551004 and RYB3064].

ORCID

Danielle Dagenais http://orcid.org/0000-0001-9071-459X
Sylvain Paquette http://orcid.org/0000-0002-7762-6525

References

Ahern, J. (2011). From fail-safe to safe-to-fail: Sustainability and resilience in the new urban world. *Landscape and Urban Planning, 100*, 341–343. doi:10.1016/j.landurbplan.2011.02.021

Ahern, J., Cilliers, S., & Niemelä, J. (2014). The concept of ecosystem services in adaptive urban planning and design: A framework for supporting innovation. *Landscape and Urban Planning, 125*, 254–259. doi:10.1016/j.landurbplan.2014.01.020

Astaraie-Imani, M., Kapelan, Z., Fu, G., & Butler, D. (2012). Assessing the combined effects of urbanisation and climate change on the river water quality in an integrated urban wastewater system in the UK. *Journal of Environmental Management, 112*, 1–9.

Barthel, S., Folke, C., & Colding, J. (2010). Social–ecological memory in urban gardens – Retaining the capacity for management of ecosystem services. *Global Environmental Change, 20*, 255–265. doi:10.1016/j.gloenvcha.2010.01.001

Bennett, E. M., Peterson, G. D., & Gordon, L. J. (2009). Understanding relationships among multiple ecosystem services. *Ecology Letters, 12*, 1394–1404. doi:10.1111/j.1461-0248.2009.01387.x

City of Quebec. (2014). *Beauport.* Retrieved from http://ville.quebec.qc.ca/apropos/portrait/quelques_chiffres/arrondissements/beauport.aspx

Clarke, L. W., Jenerette, G. D., & Davila, A. (2013). The luxury of vegetation and the legacy of tree biodiversity in Los Angeles, CA. *Landscape and Urban Planning, 116*, 48–59. doi:10.1016/j.landurbplan.2013.04.006

Cutter, S., Boruff, B., & Shirley, W. (2003). Social vulnerability to environmental hazards. *Social Science Quarterly, 84*, 242–261.

D'Ercole, R., & Metzger, P. (2009, March 31). La vulnérabilité territoriale : Une nouvelle approche des risques en milieu urbain. *Cybergeo : European Journal of Geography* [Online]. Dossiers, Vulnérabilités urbaines au sud, document 447. Retrieved from http://www.cybergeo.eu/index22022.html

Dagenais, D., Paquette, S., Fuamba, M., Servier, E.-J., Spector, A., Besson, L., & Thomas, I. (2013). Participatory design of a decision aid tool integrating social aspects for the implementation of at-source vegetated best management practices (SVBMPs) at the neighbourhood level, *Novatech 8th International Conference 2013.* Lyon, France.

Dagenais, D., Paquette, S., Fuamba, M., & Thomas-Maret, I. (2011). Keys to successful large-scale implementation of vegetated best management practices in the urban environment. *12th International Conference on Urban Drainage, International Water Association*, Porto Allegre, Brazil.

Dagenais, D., Paquette S., Thomas-Maret, I., Besson, L., Chanthalama, K., Jambon, C., … Bolduc, S. (2012). *Implantation en milieu urbain de systèmes végétalisés de contrôle à la source des eaux pluviales comme option d'adaptation aux changements climatiques.* Final report presented to OURANOS and the Programme Environnement bâti/sud du Québec (ICAR). Montreal, Canada.

Daniel, T. C., Muhar, A., Arnberger, A., Aznar, O., Boyd, J. W., Chan, K. M., … Grêt-Regamey, A. (2012). Contributions of cultural services to the ecosystem services agenda. *Proceedings of the National Academy of Sciences, 109*, 8812–8819.

Després, C., Vachon, G., & Fortin, A. (2011). Implementing transdisciplinary: Architecture and urbanism at work. In I. Doucet & N. Janssens (Eds.), *Transdisciplinary knowledge production in architecture and urbanism* (pp. 33–49). London: Springer-Science and Media BV.

Ellis, J. B. (2012). Sustainable surface water management and green infrastructure in UK urban catchment planning. *Journal of Environmental Planning and Management, 56*, 24–41. doi:10.1080/09640568.2011.648752

Entrix. (2010). *Portland's green infrastructure: Quantifying the health, energy and community livability benefits.* Final report prepared for the Bureau of Environmental Services, City of Portland, OR.

Everard, M., & McInnes, R. (2013). Systemic solutions for multi-benefit water and environmental management. *Science of the Total Environment, 461–462*, 170–179. doi:10.1016/j.scitotenv.2013.05.010

Fletcher, T. D., Shuster, W., Hunt, W. F., Ashley, R., Butler, D., Arthur, S., … Mikkelsen, P. S. (2015). SUDS, LID, BMPs, WSUD and more -The evolution and application of terminology surrounding urban drainage. *Urban Water Journal, 12*, 525–542.

Fryd, O., Backhaus, A., Jeppesen, J., Bergman, M., Toft, S. I., Birch, H., … Fratini, C. (2009, December). *Connected disconnections: Conditions for landscape-based disconnections of stormwater from the Copenhagen sewer system in the catchment area for River Harrestrup.* Working Report. The 2BG Project. Copenhagen: Technical University of Denmark, University of Copenhagen.

Fryd, O., Jensen, M. B., Ingvertsen, S. I., Jeppesen, J., & Magid, J. (2010). Doing the first loop of planning for sustainable urban drainage system retrofits: A case study from Odense, Denmark. *Urban Water Journal, 7*, 367–378.

Gill, S. F., Handley, J. F., Ennos, A. R., & Pauleit, S. (2007). Adapting cities for climate change: The role of the green infrastructure. *Built Environment, 33*, 115–133. Retrieved from http://dx.doi.org/10.2148/benv.33.1.115

Gorunescu, F. (2011). *Data mining: Concepts, models and techniques* (Vol. 12). Berlin: Springer-Verlag.

Hansen, R., & Pauleit, S. (2014). From multifunctionality to multiple ecosystem services? A conceptual framework for multifunctionality in green infrastructure planning for urban areas. *Ambio, 43*, 516–529. doi:10.1007/s13280-014-0510-2

Huang, G., Zhou, W., & Cadenasso, M. L. (2011). Is everyone hot in the city? Spatial pattern of land surface temperatures, land cover and neighborhood socioeconomic characteristics in Baltimore, MD. *Journal of Environmental Management, 92*, 1753–1759. doi:10.1016/j.jenvman.2011.02.006

Hunt, W. F., Davis, A. P., & Traver, R. G. (2012). Meeting hydrologic and water quality goals through targeted bioretention design. *Journal of Environmental Engineering, 138*, 698–707. doi:10.1061/(ASCE)EE.1943-7870.0000504

INSPQ & CERFO. (2012). *Ilots de chaleur/fraicheur urbains et température de surface.* Retrieved from http://www.donnees.gouv.qc.ca/?node=/donnees-details&id=2f4294b5-8489-4630-96a1-84da590f02ee

Keeley, M., Koburger, A., Dolowitz, D., Medearis, D., Nickel, D., & Shuster, W. (2013). Perspectives on the use of green infrastructure for stormwater management in Cleveland and Milwaukee. *Environmental Management, 51*, 1093–1108. doi:10.1007/s00267-013-0032-x

Low, S., Taplin, D., & Scheld, S. (2005). Anthropological methods for assessing cultural values. In S. Low, D. Taplin, & S. Scheld (Eds.), *Rethinking urban parks: Public space and cultural diversity* (pp. 175–210). Austin, TX: The University of Texas Press.

Maret, I., & Cadoul, T. (2008). Résilience et reconstruction durable: Que nous apprend la Nouvelle-Orléans [Resilience and rebuilding: Lessons from New Orleans]. *Annales de Géographie, 663*, 104–124. doi:10.3917/ag.663.0104

Marsalek, J., & Schreier, H. (2009). Innovative stormwater management in Canada: The way forward overview of the theme issue. *Water Quality Research Journal Canada, 44*, v–xi.

Martín-López, B., Iniesta-Arandia, I., García-Llorente, M., Palomo, I., Casado-Arzuaga, I., Amo, D. G. D., … Gómez-Baggethun, E. (2012). Uncovering ecosystem service bundles through social preferences. *PLoS ONE, 7*, e38970. doi:10.1371/journal.pone.0038970

McCall, M. K., & Dunn, C. E. (2012). Geo-information tools for participatory spatial planning: Fulfilling the criteria for 'good' governance? *Geoforum, 43*, 81–94. doi:10.1016/j.geoforum.2011.07.007

Moore, T. L. C., & Hunt, W. F. (2012). Ecosystem service provision by stormwater wetlands and ponds—A means for evaluation? *Special Issue on Stormwater in Urban Areas, 46*, 6811–6823. doi:10.1016/j.watres.2011.11.026

Ndubisi, F. (2002). *Ecological planning: A historical and comparative synthesis.* Baltimore, MD: Johns Hopkins University Press.

Novotny, V., Ahern, J., & Brown, P. (2010). Planning and design for sustainable and resilient cities: Theories, strategies, and best practices for green infrastructure. In W. Centric (Ed.), *Sustainable communities* (pp. 135–176). New York, NY: John Wiley. Retrieved from http://dx.doi.org/10.1002/9780470949962.ch3

Palomo, I., Martin-Lopez, B., Lopez-Santiago, C., & Montes, C. (2011). Participatory scenario planning for protected areas management under the ecosystem services framework: The Donana social-ecological system in Southwestern Spain. *Ecology and Society, 16*(1). Retrieved from http://www.ecologyandsociety.org/vol16/iss1/art23/

Pataki, D. E., Carreiro, M. M., Cherrier, J., Grulke, N. E., Jennings, V., Pincetl, S., … Pouyat, R. V (2011). Coupling biogeochemical cycles in urban environments: Ecosystem services, green solutions, and misconceptions. *Frontiers in Ecology and the Environment, 9*, 27–36. doi:10.1890/090220

Pyke, C., Warren, M. P., Johnson, T., LaGro, J., Jr., & Scharfenberg, J. (2011). Assessment of low impact development for managing stormwater with changing precipitation due to climate change. *Landscape and Urban Planning, 103*, 166–173.

Rivard, G. (2011). *Guide de gestion des eaux pluviales: Stratégies d'aménagement, principes de conception et pratiques de gestion optimales pour les réseaux de drainage en milieu urbain.* Québec: Ministère du Développement Durable de l'Environnement et de la Lutte contre les Changements Climatiques. Retrieved from http://mddefp.gouv.qc.ca/eau/pluviales/guide.htm

Robertson, T., & Simonsen, J. (2012). Participatory design, an introduction. In T. Robertson & J. Simonsen (Eds.), *Routledge handbook of participatory design* (pp. 1–19). New York, NY: Routledge.

Rodenburg, C. A., & Nijkamp, P. (2004). Multifunctional land use in the city: A typological overview. *Built Environment, 30*, 274–288. doi:10.2148/benv.30.4.274.57152

Setälä, H., Bardgett, R. D., Birkhofer, K., Brady, M., Byrne, L., de Ruiter, P. C., & de Vries, F. T. (2014). Urban and agricultural soils: Conflicts and trade-offs in the optimization of ecosystem services. *Urban Ecosystems, 17*, 239–253. doi:10.1007/s11252-013-0311-6

Staats, H. T. N. (2012). Chap 24 restorative environment. In S. D. Clayton (Ed.), *The Oxford handbook of environmental and conservation psychology* (pp. 445–458). Oxford: Oxford University Press.

Thomas, I., Bleau, N., Soto, A. P., Desjardin-Dutil, G., Fuamba, M., & Kadi, S. (2012). *Analyser la vulnérabilité sociétale et territoriale aux inondations en milieu urbain dans le contexte des changements climatiques, en prenant comme cas d'étude la ville de Montréal.* Rapport final déposé à OURANOS, Programme Environnement bâti, Institut d'urbanisme+Environnement bâti /sud du Québec (ICAR), Montreal, Canada.

TRCA/CVCA (Toronto and Region Conservation Authority, Credit Valley Conservation Authority). (2010). *Low impact development stormwater management planning and design guide.* Version 1. Toronto: Toronto and Region Conservation Authority, Credit Valley Conservation Authority. Retrieved from http://www.creditvalleyca.ca/low-impact-development/low-impact-development-support/stormwater-management-lid-guidance-documents/low-impact-development-stormwater-management-planning-and-design-guide/

Uzomah, V., Scholz, M., & Almuktar, S. (2014). Rapid expert tool for different professions based on estimated ecosystem variables for retrofitting of drainage systems. *Computers, Environment and Urban Systems, 44*, 1–14. doi:10.1016/j.compenvurbsys.2013.10.008

Walsh, C. J., Fletcher, T. D., & Ladson, A. R. (2005). Stream restoration in urban catchments through redesigning stormwater systems: Looking to the catchment to save the stream. *Journal of the North American Benthological Society, 24*, 690–705. doi:10.1899/04-020.1

White, I., & Alarcon, A. (2009). Planning policy, sustainable drainage and surface water management: A case study of Greater Manchester. *Built Environment, 35*, 516–530. doi:10.2148/benv.35.4.516

Wilson, C., Hunt, W., Winston, R., & Smith, P. (2015). Comparison of runoff quality and quantity from a commercial low-impact and conventional development in Raleigh, North Carolina. *Journal of Environmental Engineering, 141*, 05014005. doi:10.1061/(ASCE)EE.1943-7870.0000842

Italian stone pine forests under Rome's siege: learning from the past to protect their future

Lorenza Gasparella, Antonio Tomao ⁱᴰ, Mariagrazia Agrimi ⁱᴰ, Piermaria Corona ⁱᴰ, Luigi Portoghesi ⁱᴰ and Anna Barbati ⁱᴰ

ABSTRACT

Italian stone pine is a landmark of Mediterranean coastal areas. Today, pinewoods represent environmental amenity areas at risk, being under siege from intensive urbanisation. We present an emblematic case study in Rome's coastal strip where urban encroachment around pinewoods is somewhat overlooked by urban planning, which may be threatening for their conservation. We studied: (i) changes in land use intensification in the pinewoods' surroundings over the past 60 years (1949–2008), by means of a synthetic index of landscape conservation (ILC) ranging from 0 (maximum level of anthropogenic landscape alteration) to 1 (maximum level of landscape naturalness); (ii) influence of different landscape protection level on land use intensification. Findings show that in areas with low levels of landscape protection, the ILC had been decreasing in the first 100-m surrounding pinewoods, and within the 1-km buffer. The ILC had been rather stable within areas with high levels of landscape protection. Lessons learnt have implications for spatial development strategies to protect coastal pinewoods from external pressures due to future (planned) urban densification in their surroundings.

1. Introduction

Coastal areas represent an important asset for Europe: they harbour a rich diversity of habitats, boost the local economy (for example, tourism) and contribute to human well-being (recreation opportunities). In littoral EU Member States, coasts are also the most densely populated zones. In Spain, France and Italy, the share of the national population living within 5 km of the coastline amounts, on average, to 35% (Eurostat, 2011). The direct consequence of the expansion of human activities on the coast is that land development is advancing very rapidly to accommodate urban land uses, namely housing, tourism, transport infrastructure and coastal defences.

Coastal over-development is clearly tracked by the available data on soil sealing. The European Earth Observation Programme Copernicus estimated that in the period 2006–2009, the increase of soil sealing in littoral European countries has ranged between 2 and 13% of the coastal zone (area less than 10 km far from the coastline) (European Environment Agency [EEA], 2013).

The increasing pressure and demand for coastal resources results in a substantial reduction in the extent and functionality of terrestrial coastal ecosystems and related ecosystem services. For instance, intertidal habitats suffer from 'coastal squeeze' due to the combined effect of coastal erosion and barriers to landward migration created by coastal defences (Doody, 2013; Pontee, 2013). Mediterranean coastal pinewoods are also threatened by external pressures from land use intensification. However, their protection has attracted limited attention so far. This is the case for forests dominated by Italian stone pine (*Pinus pinea* L.) (Richardson & Rundel, 1998), a recognised landmark of many Mediterranean coastal areas (SoMF, 2013). Framed under the Mediterranean pine forest of the European Forest Types classification (Barbati, Corona, & Marchetti, 2007; Barbati, Marchetti, Chirici, & Corona, 2014), many pinewoods originate from reforestation programmes carried out since the late nineteenth century, to protect inland areas and crops against sea winds and sand deposition and to stabilise mobile dune systems. Some long established pine plantations, with undergrowth vegetation comparable to natural forest communities (*Quercetalia ilicis*), have become priority habitats under the EU Habitats Directive (2270 * Wooded dunes with *Pinus pinea* L. and/or *Pinus pinaster* Ait.).

Nowadays, the ecological value of coastal pinewoods varies widely. Coastal pinewoods can provide multiple ecosystem services (Paquette & Messier, 2010). In most sites, pine nut and timber production appear subordinate to ecological and social functions, such as biodiversity conservation, scenic beauty and recreational use (SoMF, 2013). At the same time, these pinewoods constitute 'environmental amenity areas at risk' because they are located in attractive settings for urbanisation projects (Leone & Lovreglio, 2004).

Taking a cue from the concept of 'coastal squeeze', we seek to demonstrate that Italian stone pine coastal forests have also been 'under siege' from coastal land use intensification for decades. An emblematic case study of the pinewoods in Rome's coastal strip is presented here as an example of a metropolitan area where urban encroachment around these pinewoods is somewhat overlooked by urban planning, which may be threatening for their conservation. At the same time, the maintenance and, also possibly, the expansion of forest areas is recognised as a priority for the implementation of a local green infrastructure (Barbati, Corona, Salvati, & Gasparella, 2013; Capotorti, Mollo, Zavattero, Anzellotti, & Celesti-Grapow, 2015).

The specific objectives of this paper are: (i) to analyse changes in the conservation status of the landscape surrounding pinewoods, based on a diachronic analysis of land use changes occurring over the past 60 years (period 1949–2008); (ii) to understand whether differential levels of landscape protection have influenced land use intensification in the investigated area; (iii) to highlight needs that should be addressed by local spatial development strategies to effectively ensure the conservation of coastal pinewoods in the future.

2. Materials and methods

2.1. Study area

The Italian stone pine forests analysed by this study extend over 1700 ha in the coastal zone of Rome's Metropolitan Area. The pinewoods are distributed in patches of variable size, located in Castelfusano, Ostia, and in the Castelporziano Presidential Estate, the property of the Presidency of the Italian Republic since 1948 (Figure 1).

Established between the eighteenth and mid-twentieth centuries as monocultures, with the initial goal of consolidating sand dunes, and timber and pine nut production, these plantations have evolved with time into different forest structural types due to various silvicultural treatments and natural forest dynamics (Agrimi, Bollati, Giordano, & Portoghesi, 2002; Ciancio, Travaglini, Bianchi, & Mariotti, 2009).

The first reforestation activities were carried out in Castelfusano in the eighteenth century. After the reclamation of coastal wetlands, successive waves of reforestation occurred: (i) between 1870 and 1887 (Agrimi, Bollati, & Portoghesi, 2005), (ii) in 1933—to counteract the spread of Malaria (Snowden, 2006)—and (iii) in the first decades after the end of the Second World War (Agrimi et al., 2005). In 1995,

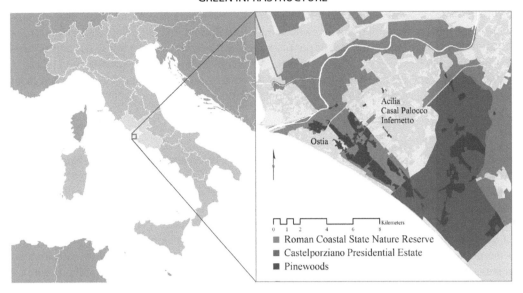

Figure 1. Maps of the study area.

the Ostia and Castelfusano pine forests were included in the 'Rome Coastal State Nature Reserve'. Four years later, the Castelporziano Presidential Estate became a designated protected area too, and a small area of about 10 ha within this estate has been recognised as priority habitat (2270 * Wooded dunes with *Pinus pinea* L. and/or *Pinus pinaster* Ait.).

The historical pinewood of Castelfusano was affected by a severe wildfire in July 2000 that destroyed some 270 ha and seriously damaged 100 ha (Minchella, Del Frate, Capogna, Anselmi, & Manes, 2009). Reforestation with *P. pinea* has been carried out in burned areas.

Because of the socio-economic changes that occurred in the past decades, these forests have lost their productive function, while gaining new values for the well-being of urban dwellers and outdoor recreation (Agrimi et al., 2005; Carrus et al., 2015). Nowadays pinewood conservation must confront two main challenges: external land use pressures and natural forest dynamics.

Preserved by the tillage technique, consisting of alternating stripes of tilled soil with bands of natural vegetation (the so-called *Pavari technique*) (Agrimi et al., 2002), many species of the pre-existing natural vegetation (*Quercus ilex* L., *Arbutus unedo* L., *Erica arborea* L., *Pistacia lentiscus* L., *Phillyrea latifolia* L., *Myrtus communis* L.) grow under the pine cover today. For example, the oldest tract of Castelfusano pine forest has developed, over time, into a complex structure due to the effect of natural dynamics and silvicultural treatments, mainly selective fellings (Agrimi et al., 2002; Ciancio, Cutini, Mercurio, & Veracini, 1986). In 2002, the youngest part of the pinewood was selectively thinned, to reduce its high density and to improve ecosystem resilience (Amorini, Cutini, & Manetti, 2002).

The pine forests of the State Natural Reserve of Castelporziano consist mostly of even-aged stands, (covering 85% of the natural reserve) with a large area of mature stands; the strategies of the current forest management plan are directed mainly towards obtaining the natural regeneration of Italian stone pine.

2.2. Land use history

The Rome coastal landscape has been affected by significant changes over the past 140 years. Urban development expanded towards the coastal zone after the construction of the Rome–Lido railway line, and the road axes 'via del Mare' and 'via Cristoforo Colombo'. After the Second World War, urban encroachment accelerated, even leading to the illegal construction of entire neighbourhoods. The 1933 and 1936 Rome masterplans also triggered residential expansion in the coastal areas. The population living in the areas surrounding the pinewoods increased tenfold from 33 949 to 331 918 in 40 years (1950–1990), and the maximum rate of population growth occurred during the sixties and seventies (Duchini & Tinelli, 1996; Tinelli & Bellini, 1997). This complex development process has created the two urban areas of Ostia and Acilia-Casal Palocco-Infernetto, separated by large unbuilt areas mainly represented by pinewoods (see Figure 1) (Coppola, Fausti, & Romualdi, 1997; Creti, 2008).

2.3. Data-set

The diachronic analysis of land use changes occurring in the landscape surrounding the pinewoods over the past 60 years was performed using four available 1:25 000 scale digital land cover maps: the 1949 map (LCM49), derived from a set of land use maps originally devised by the Italian Military Geographical Institute (IGM) and vectorised in a previous project (Salvati, 2013); the 'Agricultural and forest map of Rome region', produced by the district authority of Rome in 1974 (LCM74); the 1999 land cover map (LCM99) devised by the Regional Authority of Latium (minimum mapping unit of 1 ha) from digital ortho-photo interpretation (Terraitaly - IT2000, 1998–1999, 1 m pixel); and the 2008 map (LCM08), a high-resolution (.5 m) land cover map processed from photo-interpretation of digital ortho-images released from the Italian National Geoportal (Italian Ministry for Environment, Land and Sea).

All original land cover maps were reclassified into eight broad Corine-like land cover types, plus a ninth one which includes only pinewoods, since the land cover classes used in the four maps are not fully comparable (see Table 1). The adopted land cover classification system is based on a minimum mapping unit of 1 hectare. Pinewood polygon boundaries (2008) were also validated against aerial images captured by Terraltaly™ in 2008, with the support of ancillary information (Agrimi et al., 2002).

2.4. Land use change analysis

The geodatasets LCM49, LCM74, LCM99 and LCM08 were used for a multi-temporal assessment of the proportion of each land cover class within the Total Area (TA) of multiple-ring buffers created at increasing distances (100, 250, 500, 1000 m) around pinewoods (Figure 2).

This method allowed us to assess pressures in the areas close to pinewoods (100 m) that can determine negative edge effects. The edge effect is the result of a series of physical, chemical and ecological interactions in the contact area between adjacent habitat types separated by an abrupt transition zone. For example, the longer the common border between forest or shrubland and urban settlements, the higher the risk of wildfires that, in turn, might be a threat for the population of surrounding urban areas (see, for example, Lafortezza et al., 2015).

The 1-km buffer allows a wider swathe of the surrounding landscape to be analysed, taking into account the different legal constraints acting therein. The intermediate 250 and 500-m buffers were intended to assess whether land use change follows any spatial patterns, at increasing distances from pinewoods.

Table 1. ILC coefficients assigned to land use/cover classes characterising the landscape surrounding pinewoods.

Corine classes	ILC coefficient
1.1 Urban fabric	1
1.4 Artificial, non-agricultural, vegetated areas	2
2.1 Arable land	3
2.2 Permanent crops	4
2.3 Pastures	7
2.4 Heterogeneous agricultural areas	3
3.1 Forests	8
5.1 Inland waters	7

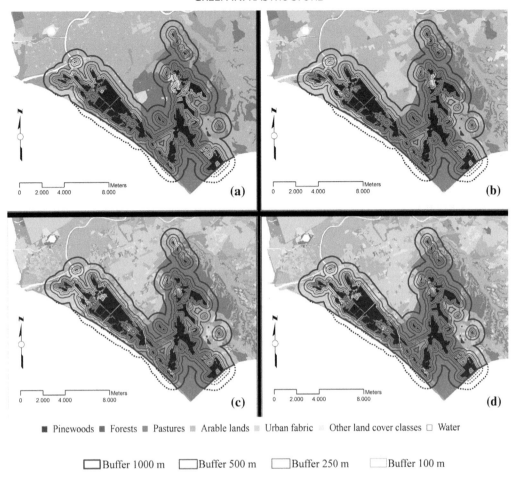

Figure 2. Buffers at 100, 250, 500, 1000 m around the pinewoods over land cover maps at the years 1949 (a), 1974 (b), 1999 (c) and 2008 (d). The dashed line represents the buffers in no-data areas (the seaside and areas outside Rome's municipality). Artificial non-agricultural vegetated areas, permanent crops, heterogeneous agricultural areas and inland waters are displayed as 'other land cover classes', since they cover altogether only 5% of the TA.

Variations in the level of land use intensification observed over the past sixty years were also evaluated by means of the Index of Landscape Conservation (ILC). Originally proposed by Pizzolotto and Brandmayr (1996), and recently applied in setting landscape restoration priorities in the metropolitan area of Rome (Capotorti et al., 2015), the ILC quantifies the landscape conservation condition of a given area by assigning each land use/land cover class to ordinal values (Table 1), defined in relation to levels of anthropogenic alteration of the natural environment (e.g., soil sealing, hemeroby). The higher the land use intensification (e.g., urban areas, arable lands), the lower the ordinal value. The ILC is then calculated as follows:

$$ILC = \sum_i^n \frac{sur_i}{sur} \times \frac{coeff_i}{coeff_n}$$

where n is the number of ILC classes identified for the study area, sur_i is the surface area of the i-th ILC class; sur is the TA of the spatial unit enclosed by each multiple-ring buffer; $coeff_i$ is the coefficient assigned to the i-th ILC class; and $coeff_n$ is the maximum value of the ILC in the study area.

ILC scores range from 0 (maximum relative level of landscape anthropogenic alteration) to 1 (maximum relative level of landscape naturalness).

Both land use and ILC changes were assessed with respect to the level of landscape protection of the territory, distinguishing, in the TA of each single buffer, a High Level of Landscape Protection Area (HLLPA) from Low Level of Landscape Protection Area (LLLPA). The HLLPA is the Presidential Estate of Castelporziano where building is forbidden, access is restricted and there is strict surveillance of the permitted activities. Conversely, the LLLPA includes areas designated for urban development by the town masterplan. Building is also allowed as part of the redevelopment and densification of illegally built settlements.

3. Results

3.1. Land use change around pinewoods (1949–2008)

The total analysed buffer areas range from 1535 (100-m buffer) to 8440 (1000-m buffer) hectares. The land use change that has been occurring in the buffer areas surrounding pinewoods from 1949 to 2008 is summarised in Table 2. Covering about 95% of the TA, 'forests', 'pastures', 'arable land' and 'urban fabric' are the land use classes most involved in landscape transformation. In 1949, urban areas occupied 3.2% of the TA within the 1-km buffer, whereas agricultural and forested land covered 28.3 and 46.8%, respectively. In 2008, urban areas peaked at close to 15%, whereas agricultural lands dropped to 18.8% of the TA. A moderate increase in forest land has been occurring over the same period. However, the positive trend of forest land increase, observed in the TA of all buffers, is the net effect of forest expansion in the HLLPA, and forest area losses in the LLLPA.

On one hand, forests have been naturally expanding within HLLPA, at the expense of agricultural land. On the other, forest and agricultural land have been squeezed within the LLLPA, due to urban growth. The high rate of urban encroachment within the LLLPA is reflected by the strikingly high share of urban fabric surrounding pinewoods in 2008, ranging from 16% (buffer 100 m) to 30% (buffer 1 km).

The land use change analysis shows that the rate of change was not consistent over the considered period. A higher rate of cover change characterised the 1949–1974 and 1974–1999 periods. A stabilisation occurred in the final decade (1999–2008) as regards forests. The annual change percentages of the forest in the LLLPA were respectively –.23 and –.11% in the first two periods and stabilised in the last period (+.03%). Urban fabric is the most dynamic land cover class. The annual percentage of change was +.44 and +.3% in the 1949–1974 and 1974–1999 periods, respectively, while it has been characterised by a moderate total increase in the 1999–2008 period (+.58%).

3.2. Landscape conservation around pinewoods

The direction and magnitude of changes in the quality of the landscape surrounding pinewoods is highlighted by the ILC findings (Figures 3 and 4). From 1949 to 2008, the ILC was stable at around .7 in the TA of the 500-m and 1000-m buffers, while slightly increasing in the 100-m and 250-m buffers. Again, these trends result from the offsetting of the ILC decline within LLLPA. In fact, the ILC had been decreasing by approximately .1 in all buffer areas within the LLLPA: ranging from .71 in 1949 to .63 in 2008 in the first 100 m surrounding pinewoods and from .54 to .41 within the 1-km buffer. On the other hand, the ILC had been stable at around .8 (or slightly increasing) within the HLLPA.

While the ILC temporal trend is similar across all the buffer areas, its spatial variation is distance-dependent, at particular times, in the LLLPA (Figure 4). Landscape surroundings at a distance of less than 250 m from pinewoods were in a better landscape conservation condition compared to wider buffers during the whole period. Spatial variation in the ILC was less pronounced in 1949 and became evident from 1974 in the LLLPA. The ILC downward trend at increasing distance from pinewoods is observed considering all the buffers of TA, although to a lesser degree. Again this results from the offsetting

Table 2. Temporal trend of land cover classes surrounding pinewoods with respect to TA (first block), HLLPA (second block) and LLLPA (third block).

TA

Corine classes	Buffer 100 m				Buffer 250 m				Buffer 500 m				Buffer 1000 m			
	1949	1974	1999	2008	1949	1974	1999	2008	1949	1974	1999	2008	1949	1974	1999	2008
1.1 Urban fabric	1.1	2.7	5.6	7.0	1.3	4.7	7.4	8.9	1.9	7.4	9.2	11.4	3.2	8.6	12.6	15.2
1.4 Artificial, non-agricultural, vegetated areas	1.0	2.4	3.4	2.0	1.4	2.3	3.4	2.0	1.7	1.8	4.1	2.7	1.2	1.2	3.5	2.6
2.1 Arable land	22.3	8.9	6.3	7.3	22.7	13.2	9.1	9.8	23.8	20.0	13.9	14.1	28.3	28.2	18.8	18.8
2.2 Permanent crops	.0	.0	.0	.0	.0	.0	.0	.0	.0	.0	.1	.0	.1	.0	.2	.2
2.3 Pastures	12.4	16.4	16.2	10.2	16.1	14.9	18.5	11.9	19.5	12.5	18.0	12.3	17.9	10.9	16.3	11.1
2.4 Heterogeneous agricultural areas	2.2	.0	.0	.1	2.4	.0	.0	.3	2.1	.0	.2	.9	1.7	.1	.6	1.3
3.1 Forests	60.9	69.4	67.9	72.6	55.8	64.8	61.0	66.4	50.8	57.7	54.0	58.1	46.8	50.2	47.3	50.0
5.1 Inland waters	.1	.0	.8	.8	.2	.1	.6	.6	.3	.5	.6	.6	.7	.8	.8	.8

HLLPA

Corine classes	Buffer 100 m				Buffer 250 m				Buffer 500 m				Buffer 1000 m			
	1949	1974	1999	2008	1949	1974	1999	2008	1949	1974	1999	2008	1949	1974	1999	2008
1.1 Urban fabric	.3	.0	.6	.6	.2	.0	.7	.7	.1	.0	.4	.4	.1	.0	.3	.3
1.4 Artificial, non-agricultural, vegetated areas	.6	2.5	.0	.0	.8	2.0	.0	.0	.5	1.3	.0	.0	.4	.9	.0	.0
2.1 Arable land	21.8	3.8	1.9	1.2	16.9	4.1	3.1	2.2	12.6	6.0	5.5	4.6	10.7	5.9	5.8	5.0
2.2 Permanent crops	.0	.0	.0	.0	.0	.0	.0	.0	.0	.0	.0	.0	.0	.0	.0	.0
2.3 Pastures	11.9	16.6	18.9	12.1	14.9	15.2	18.8	11.6	16.8	13.0	16.7	10.8	16.1	10.4	14.0	9.7
2.4 Heterogeneous agricultural areas	3.1	.0	.0	.1	3.4	.0	.0	.2	3.1	.0	.0	.1	2.5	.0	.0	.1
3.1 Forests	62.3	77.2	77.6	85.0	63.8	78.7	76.8	84.7	66.8	79.6	76.9	83.5	70.2	82.4	79.5	84.5
5.1 Inland waters	.0	.0	1.0	1.0	.0	.0	.6	.6	.0	.1	.5	.5	.0	.3	.4	.4

LLLPA

Corine classes	Buffer 100 m				Buffer 250 m				Buffer 500 m				Buffer 1000 m			
	1949	1974	1999	2008	1949	1974	1999	2008	1949	1974	1999	2008	1949	1974	1999	2008
1.1 Urban fabric	2.2	6.7	12.7	16.2	3.1	12.0	17.6	21.4	4.4	17.4	21.1	26.2	6.3	17.2	24.7	30.0
1.4 Artificial, non-agricultural, vegetated areas	1.6	2.4	8.2	4.8	2.4	2.7	8.7	5.2	3.2	2.6	9.6	6.3	2.1	1.5	6.9	5.1
2.1 Arable land	23.0	16.3	12.6	16.0	31.7	26.9	18.2	21.4	38.9	39.1	25.3	27.0	45.7	50.2	31.6	32.5
2.2 Permanent crops	.0	.0	.0	.0	.0	.0	.0	.0	.0	.0	.1	.0	.3	.0	.5	.3
2.3 Pastures	13.3	16.3	12.2	7.5	17.9	14.4	18.0	12.4	23.1	11.8	19.7	14.2	19.6	11.4	18.5	12.5
2.4 Heterogeneous agricultural areas	1.0	.0	.0	.1	1.0	.0	.0	.5	.7	.1	.4	1.9	.8	.2	1.1	2.6
3.1 Forests	58.7	58.3	53.9	54.8	43.5	43.7	36.8	38.5	29.1	28.0	23.0	23.6	23.8	18.2	15.5	15.8
5.1 Inland waters	.2	.1	.5	.5	.4	.3	.7	.7	.6	1.0	.8	.8	1.4	1.4	1.2	1.2

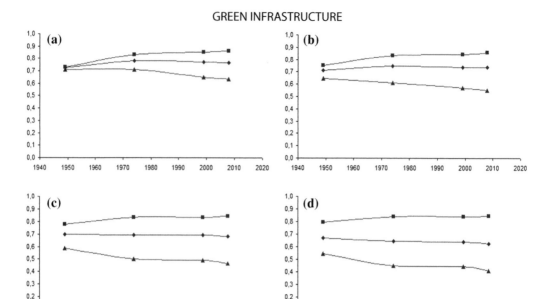

Figure 3. Temporal trend of ILC at 100 m (a), 250 m (b), 500 m (c), 1000 m (d) buffer distances.

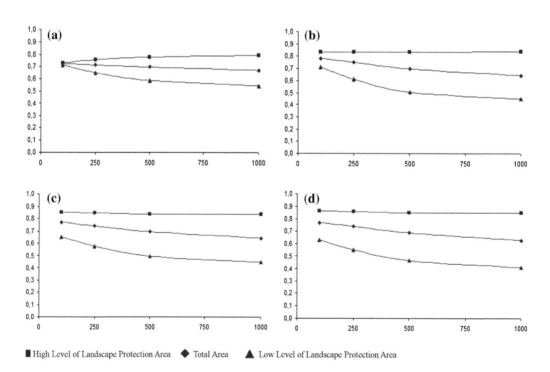

Figure 4. ILC values across buffer distances in 1949 (a), 1974 (b), 1999 (c), 2008 (d).

effect between differential ILC patterns occurring in the LLLPA as opposed to the HLLPA, where the ILC appears steady across space.

4. Discussion

Buffer analysis has been successfully applied to evaluate landscape change around protected areas (Kintz, Young, & Crews-Meyer, 2006) and cities (Seto & Fragkias, 2005). It gives relevant information to landscape and urban planners, such as the distances around residual elements of the natural landscape that should be specifically preserved. In our study, buffer analysis clearly demonstrates that Rome's urban encroachment has been gradually putting Italian stone pine coastal forests 'under siege' over the past 60 years. On the other hand, the strict protection offered by the designation of the area of Castelporziano has worked as an effective shield to protect much of the landscape around pinewoods.

In 2008, the TA of the buffer ring closest to the pinewoods (100 m) retained a huge proportion of rural land consisting of forests (73%), pastures (10%) and agricultural areas (7%). Nevertheless, there are remarkable differences between the HLLPA, where forests covered almost all the buffer area, and the LLLPA, where forests covered 55% only, and urban areas amounted to 16%. While the proportion of forest and urban areas was relatively even across HLLPA buffer rings in 2008 (84–85% and .3–.7% respectively), an opposite trend has been recorded in the LLLPA. Forest share dropped from 55% in the buffer ring closest to the pinewoods to just 16% within the 1-km buffer. Conversely, the 30% urban areas share within the 1-km buffer is almost double the value observed in the immediate vicinity of pinewoods.

As a consequence of differential patterns of land use intensification, ILC values are characterised by divergent trends between HLLPA (stable pattern between buffer rings, slightly increasing over time) and LLLPA (decreasing from 100-m to 1-km ring, with a sharp decline in areas at a distance of more than 250 m from pinewoods, and decreasing over time). In fact, the Castelporziano Presidential Estate has gradually implemented measures to protect natural ecosystems and crops, historically managed to yield profits, and nowadays regarded as elements of landscape heterogeneity to be preserved. Thus, the assessment of external pressures threatening the conservation of a given natural habitat can lead to partial or misleading results, if it does not properly consider the local effects of land management and legal constraints (see also, Antrop, 1998, 2005), as demonstrated by our investigation across space (at increasing distances from a given natural habitat) and time (diachronic analysis).

What we have learnt from past landscape dynamics has implications for spatial development strategies aimed at protecting coastal pinewoods from external pressures due to future (planned) urban densification. To this end, the establishment of safeguard measures to protect residual arable lands and pastures surrounding pinewoods from further urban encroachment would prevent a likely increase of nearby anthropogenic pressures on pinewoods, such as fragmentation, pollution and expansion of the wildland–urban interface and related risk of wildfire (Lafortezza et al., 2015; Leone & Lovreglio, 2004). Reducing risk of wildfire initiation, using silvicultural measures (Corona et al., 2015), is a high priority for the conservation of Italian stone pine forests. The high level of flammability of these thermophilous pinewoods (Corona, Ferrari, Cartisano, & Barbati, 2014) combined with the ecological traits of *P. pinea*, a non-serotinous species, makes successional trajectories after fire disturbances that are most likely to evolve towards other vegetation types (Retana et al., 2012), unless active restoration is implemented.

The recently enforced Rome's masterplan has introduced the ecological network as a legally binding instrument to protect a wide area of Rome's Municipality (Rome's masterplan, 2003) from damaging landscape transformations. This green infrastructure amounts to 67% of Rome's Municipal area and includes protected areas, agricultural lands and green urban areas (Capotorti et al., 2015). Whilst coastal pinewoods are included as primary components of the green infrastructure and a high level of landscape protection is envisaged in the buffer zone within 100 m from the Castelporziano estate, no protection measures are provided around the other patches of coastal pinewood. On the contrary, urban planning permits urban densification in areas (arable lands and pastures) close to pinewoods, namely the redevelopment of existing illegal buildings, and also the establishment of new settlements. Even the agricultural areas included within the 'Rome Coastal State Nature Reserve' are threatened by urban encroachment. The management plan of the Reserve, still not approved by the Lazio Region (the territorial administrative body at sub-national level), identifies protection measures for agricultural areas, but, strangely enough, allows the establishment of new buildings in the protected area.

5. Conclusion

Today, the Castelfusano, Ostia and Castelporziano coastal pinewoods represent important hubs of Rome's green infrastructure. Findings from our retrospective analysis showed that, despite intense urbanisation, there is still a good degree of landscape conservation in the immediate surroundings of these forest areas. This condition must be actively preserved in future, if coastal pinewoods are to be retained. *Pinus pinea* forest conservation calls for an effective blend of dynamic conservation, to enable forest stands to evolve and, therefore, to preserve their genetic variation under conditions of environmental change, as well as spatial planning.

On one hand, the management of coastal pinewoods, based on a participatory approach, should shift from the classical even-aged forest for timber and pine nut production to silvicultural treatments based on selective cutting, capable of triggering natural regeneration.

On the other hand, as demonstrated by the case study, spatial planning can reduce the imbalances in land use intensification in the Metropolitan area of Rome only by taking into account the complexity of the regulatory framework (legal environmental protection framework, urban planning) affecting landscape transformations. Despite the conversion of forest into other land uses being prohibited by law in Italy, such a legal constraint does not impose a limit on external land use pressures around forest margins. In addition, the inclusion of coastal pinewoods as primary components of the local green infrastructure must not be used as the justification for land use intensification in the surrounding landscape. The maintenance of a buffer of rural land uses around pinewoods seems the most viable option for protecting forests from external pressures. In fact, rural land around pinewoods, which can accommodate dynamic change, is likely to protect coastal pinewoods by providing more resilience to disturbances brought about by coastal urbanisation.

Acknowledgements

The authors are grateful for the suggestions of the anonymous reviewers.

Disclosure statement

No potential conflict of interest was reported by the authors.

Funding

This work was supported by the project 'Innovative models for the analysis of ecosystem services of forests in urban and periurban context – NEUFOR' (PRIN 2012 - 2012K3A2HJ) funded by the Italian Ministry of University and Research (project coordinator: G. Sanesi). The research was also carried out within the COST Action FP 1204 'Green Infrastructure approach: linking environmental with social aspects in studying and managing urban forests' (chair: C. Calfapietra).

ORCiD

Antonio Tomao http://orcid.org/0000-0001-6656-400X
Mariagrazia Agrimi http://orcid.org/0000-0001-6310-4892
Piermaria Corona http://orcid.org/0000-0002-8105-0792
Luigi Portoghesi http://orcid.org/0000-0002-8964-3779
Anna Barbati http://orcid.org/0000-0002-9064-0903

References

Agrimi, M., Bollati, S., Giordano, E., & Portoghesi, L. (2002). Struttura dei popolamenti e proposte di gestione per le pinete del litorale romano [Stand structure and management proposals for Italian stone pine forests of the Roman coastline]. *Italian Journal of Forest and Mountain Environments, 57*, 242–258. Retrieved from http://ojs.aisf.it/index.php/ifm/article/view/705/675 [in Italian].

Agrimi, M., Bollati, S., & Portoghesi, L. (2005). Funzioni e valori ambientali, sociali e culturali delle pinete di Castelfusano e Ostia [Environmental, social and cultural functions and values of the Castelfusano and Ostia Italian stone pine forests]. *Proceedings of conference 'Ecosistema Roma'*, April 14–16, Roma: Atti dei Convegni Lincei, *218*, 449–457. Italian National Academy of Lincei. [in Italian].

Amorini, E., Cutini, A., & Manetti, M. C. (2002). Indagini ecologico-strutturali e indicazioni per la gestione selvicolturale [Ecological and structural surveys for forest management]. In C. Blasi & B. Cignini (Eds.), *Il recupero ambientale della pineta di Castel Fusano: studi e monitoraggi* [The environmental recovery of Castel Fusano] (pp. 24–28). Rome: Palombi Editori. [in Italian].

Antrop, M. (1998). Landscape change: Plan or chaos? *Landscape and Urban Planning, 41*, 155–161.

Antrop, M. (2005). Why landscapes of the past are important for the future. *Landscape and Urban Planning, 70*, 21–34.

Barbati, A., Corona, P., & Marchetti, M. (2007). A forest typology for monitoring sustainable forest ecosystem management: The case of European forest types. *Plant Biosystems, 1*, 93–103.

Barbati, A., Corona, P., Salvati, L., & Gasparella, L. (2013). Natural forest expansion into suburban countryside: Gained ground for a green infrastructure? *Urban Forestry & Urban Greening, 12*, 36–43.

Barbati, A., Marchetti, M., Chirici, G., & Corona, P. (2014). European forest types and forest Europe SFM indicators: Tools for monitoring progress on forest biodiversity conservation. *Forest Ecology and Management, 321*, 145–157.

Capotorti, G., Mollo, B., Zavattero, L., Anzellotti, I., & Celesti-Grapow, L. (2015). Setting priorities for urban forest planning. A comprehensive response to ecological and social needs for the metropolitan area of Rome (Italy). *Sustainability, 7*, 3958–3976.

Carrus, G., Scopelliti, M., Lafortezza, R., Colangelo, G., Ferrini, F., Salbitano, F., … Sanesi, G. (2015). Go greener, feel better? The positive effects of biodiversity on the well-being of individuals visiting urban and peri-urban green areas. *Landscape and Urban Planning, 134*, 221–228.

Ciancio, O., Cutini, A., Mercurio, R., & Veracini, A. (1986). Sulla struttura della pineta di pino domestico di Alberese [On the Alberese Stone pine (*Pinus pinea* L) stand structure]. *Annali dell'Istituto Sperimentale per la Selvicoltura, 17*, 169–236. [in Italian].

Ciancio, O., Travaglini, D., Bianchi, L., & Mariotti, B. (2009). La gestione delle pinete litoranee di pino domestico: il caso dei 'Tomboli di Cecina' [The management of Coastal Stone Pine Forests: The case of 'Tomboli di Cecina']. *Proceedings of the III National congress of Silviculture* (Vol. 1, pp. 156–162). Florence: Italian Accademy of Forest Sciences. [in Italian].

Coppola, F., Fausti, G., & Romualdi, T. (1997). *La città interrotta — Ostia Marittima 1904–1944* [The unfinished town - Ostia marittima 1904–1944]. Ostia: Edizioni Centro Studi Sinesi. [in Italian].

Corona, P., Ascoli, D., Barbati, A., Bovio, G., Colangelo, G., Elia, M., … Chianucci, F. (2015). Integrated forest management to prevent wildfires under Mediterranean environments. *Annals of Silvicultural Research, 39*, 1–22.

Corona, P., Ferrari, B., Cartisano, R., & Barbati, A. (2014). Calibration assessment of forest flammability potential in Italy. *iForest-Biogeosciences and Forestry, 7*, 300–305.

Creti, L. (2008). *Il Lido di Ostia* [Ostia Seaside] (pp. 21–26). Rome: Libreria dello Stato, Istituto Poligrafico e Zecca dello Stato. [in Italian].

Doody, J. P. (2013). Coastal squeeze and managed realignment in southeast England, does it tell us anything about the future? *Ocean & Coastal Management, 79*, 34–41.

Duchini, D., & Tinelli, A. (1996). Rilevamento ed analisi dei dati relativi all'abusivismo edilizio [Survey data and analysis of illegal building]. In *Progetto di Monitoraggio Ambientale della Tenuta presidenziale di Castelporziano*. Rapporto 1996. [Project of environmental monitoring of Castelporziano presidential Estate. Report 1996] (pp. 263–269). Rome: SITAC (Sistema Informativo Territoriale Ambientale Castelporziano). [in Italian].

EEA. (2013). *Balancing the future of Europe's coasts — Knowledge base for integrated management* (Report No 12/2013). Copenhagen: Author. [Adobe Digital Editions version]. Retrieved from http://www.eea.europa.eu/publications/balancing-the-future-of-europes

Eurostat. (2011). *Coastal regions: People living along the coastline, integration of NUTS 2010 and latest population grid*. Statistics in focus 30/2013. Retrieved from http://ec.europa.eu/eurostat/statistics-explained/index.php/Coastal_regions_-_population_statistics

Kintz, D. B., Young, K. R., & Crews-Meyer, K. A. (2006). Implications of land use/land cover change in the buffer zone of a National Park in the Tropical Andes. *Environmental Management, 38*, 238–252.

Lafortezza, R., Tanentzap, A. J., Elia, M., John, R., Sanesi, G., & Chen, J. (2015). Prioritizing fuel management in urban interfaces threatened by wildfires. *Ecological Indicators, 48*, 342–347.

Leone, V., & Lovreglio, R. (2004). Conservation of Mediterranean pine woodlands: Scenarios and legislative tools. *Plant Ecology (formerly Vegetatio), 171*, 221–235.

Minchella, A., Del Frate, F., Capogna, F., Anselmi, S., & Manes, F. (2009). Use of multitemporal SAR data for monitoring vegetation recovery of Mediterranean burned areas. *Remote Sensing of Environment, 113*, 588–597.

Paquette, A., & Messier, C. (2010). The role of plantations in managing the world's forests in the *Anthropocene. Frontiers in Ecology and the Environment, 8*, 27–34. doi:10.1890/080116

Pizzolotto, R., & Brandmayr, P. (1996). An index evaluate landscape conservation state based on land-use pattern analysis and geographic information system techniques. *Coenoses, 11*, 37–44.

Pontee, N. (2013). Defining coastal squeeze: A discussion. *Ocean & Coastal Management, 84*, 204–207.

Retana, J., Arnan, X., Arianoutsou, M., Barbati, A., Kazanis, D., & Rodrigo, A. (2012). Non serotinous pine forests. Chapter 7. In F. Moreira, M. Arianoutsou, P. Corona, & J. De las Heras (Eds.), *Post-fire management and restoration of Southern European forests* (pp. 151–170). New York, NY: Springer. ISBN: 978-94-007-2207-1. doi: 10.1007/978-94-007-2208-8_7

Richardson, D. M., & Rundel, P. W. (1998). Ecology and biogeography of *Pinus*: An introduction. In D. M. Richardson (Ed.), *Ecology and biogeography of Pinus* (pp. 3–46). Cambridge: Cambridge University Press.

Rome's masterplan. (2003). Retrieved from http://www.urbanistica.comune.roma.it/prg-adottato-4.html

Salvati, L. (2013). Monitoring high-quality soil consumption driven by urban pressure in a growing city (Rome, Italy). *Cities, 31*, 349–356.

Seto, K. C., & Fragkias, M. (2005). Quantifying spatiotemporal patterns of urban land-use change in four cities of China with time series landscape metrics. *Landscape Ecology, 20*, 871–888.

Snowden, F. M. (2006). *The conquest of malaria: Italy, 1900–1962*. New Haven, CT: Yale University Press.

SoMF. (2013). *State of Mediterranean forests*. FAO 2013. Forestry Department. [Adobe Digital Editions version]. Retrived from http://www.fao.org/forestry/silvamed/35347/en/

Tinelli, A., & Bellini, A. (1997). Analisi dell'evoluzione della popolazione e previsione dello sviluppo demografico [Population growth analysis and demographic trend forecasting]. In *Progetto di monitoraggio ambientale della Tenuta Presidenziale di Castelporziano, gruppo di lavoro Impatto antropico*. Rapporto 1997 [Project of environmental monitoring of Castelporziano presidential Estate. Report 1997] (pp. 316–333). Rome: SITAC (Sistema Informativo Territoriale Ambientale Castelporziano). [in Italian].

Defining community-scale green infrastructure

Gemma Jerome

ABSTRACT

Over the last 15 years, we have seen green infrastructure planning develop and refine its focus. The observable shift is from a focus on *what*, to *why* and more recently, *how* we deliver green infrastructure. In the urban context, there is often an emphasis on the capacity of *strategic* level projects to deliver the plurality of functions and benefits we have come to expect from our towns and cities. However, PhD research conducted at the University of Liverpool brings into focus the potential for small scale green infrastructure sites to respond to green infrastructure needs. As such a new concept of *community*-scale green infrastructure is introduced to describe activity at the local level. With reference to examples from research in The Mersey Forest Community Forest area of the north-west of England, community-scale green infrastructure is understood as a network of groups and projects who aim to deliver locally relevant functions and benefits to respond effectively to changing social and environmental needs.

Introduction

Broadly speaking, current academic thinking allows analysis of small scale voluntary activity within green infrastructure through two main lenses. Firstly, small scale green infrastructure sites may be analysed in terms of social outcomes. For example, the dynamics between people and place, and the recent popularity of community gardens as a site of social capacity building (Firth, Maye, & Pearson, 2011; Johnson, 2012; Zoellner, Zanko, Price, Bonner, & Hill, 2012). These studies primarily measure outcomes utilising qualitative social indicators such as personal well-being, confidence and self-esteem; and collective pride and cohesion at the local level. This approach is typified by Firth et al.'s study which found that access to a community garden encouraged 'a greater sense of pride and motivation to make aesthetic changes' to small scale sites of green infrastructure (2011, p. 557). Secondly, relating to the physical and ecological attributes of green infrastructure, small scale voluntary activity may be analysed in terms of the contribution it makes to ecological networks, and specifically how it improves instances of 'interconnectivity' between sites of green infrastructure (Benedict & McMahon, 2006) and 'continuity' across 'hubs' (larger areas) and 'sites' (smaller areas) (Ahern, 2007, p. 241). As such, small scale green infrastructure becomes a significant mechanism for delivering multi-functional benefits attributed to green infrastructure more generally, and supports Natural England's understanding of a 'multi-scale' approach to delivery (2009, p. 9). This paper is principally concerned with the first focus and presents new insights into the role of small scale green infrastructure plays in contributing social outcomes.

Defining community-scale green infrastructure

In order to define the specifics of community-scale green infrastructure, it is first necessary to establish the parameters by which green infrastructure in general is understood. It may be reasoned that, over the past 15 years, the advocacy argument for delivering multi-functional green and blue spaces through the framework of green infrastructure planning has largely been won (Benedict & McMahon, 2006; Mazza et al., 2011; National Research Council, 2004). Attention has consequently turned to the effectiveness of existing and emergent mechanisms which successfully deliver *function/s* in response to an area of identified green infrastructure *need,* with the aim of delivering *benefit/s* to human or non-human actors (The Mersey Forest, 2014, p. 8). In the context of implementation, particularly at the European level of green infrastructure guidance, *function* and *service* are used interchangeably. Examples of function within the environmental service vernacular include carbon storage and sequestration, water purification, air quality, and production of food, fibre and fuel (EC European Commission, 2012). In simple terms, *function* refers to the question of 'how' in green infrastructure.

In contrast, *benefit* refers to the question of 'what' and 'why'; and details the quality and quantity of value transferred to a range of beneficiaries, for example 'increased yield attributable to soil quality,' 'perception of the attractiveness of an area for workers/investors,' or 'number of visitors per year' (EC European Commission, 2012; The Mersey Forest, 2012, 2014). Green infrastructure *need* relates to both function and benefit, such as density of tree cover in a neighbourhood in need of urban cooling due to high seasonal temperatures; and describes the 'quality, distance and quantity' of a site of green infrastructure (CABE Space, 2009). In addition, *need* may also refer to addressing the exclusion of a certain social group/s to these benefits, and therefore highlights the role green infrastructure provision can play in enhancing quality of life through sustainable land management approaches (The Mersey Forest, 2013, p. 29). As a consequence, when analysing the contribution of green infrastructure elements within a specific geo-spatial context, it is possible to measure *function, benefit* or *need* or a combination of all three.

Beyond the variables of *function, benefit* and *need,* green infrastructure can be defined by its *scale.* The Landscape Institute (2013) emphasises the importance of the landscape scale when considering how to integrate green infrastructure within urban developments. Similarly, Natural England and Land Use Consultants (2009) iterate the role of green infrastructure in creating a framework for delivering large-scale environmental improvements across urban and rural contexts, highlighting the role of 'ecological networks' and 'green corridors' which intersect local authority boundaries necessitating joined up approaches to management (2009, p. 8). Within the same guidance document, Natural England references green infrastructure as a 'strategic, multi-scale' approach to 'land conservation and land use planning' (2009, p. 9). Furthermore, the strategic role of green infrastructure in creating both a conceptual and practical framework for sustainable land management is evidenced in the emergent number of green infrastructure frameworks for both urban and rural areas (e.g. Liverpool City Region and Warrington, 2013 and 2014; North East Wales, Cheshire and Wirral, 2010; Greater Manchester, 2008; South East, 2009). As such, it would be feasible to interpret coalescence of different stakeholder groups around a shared understanding that for maximal function/s and benefit/s, the preferred scale of green infrastructure delivery and enhancement is the landscape scale or the *strategic scale.* Working at such a *strategic scale* has recently been typified by large-scale projects with a landscape scale or technology focus, exemplified by projects such as 'Wirral Waters' (Peel Land & Holdings, 2015) in Merseyside which utilises green infrastructure as the context for attracting investment to redevelop an assemblage of brownfield sites. This type of green infrastructure delivery may be thought of as exemplifying the 'business case for green infrastructure' (Alker, 2015), reflecting the potential to converge green infrastructure development with significant capital investment. However, the last five years have also seen a rise in academic literature on the multi-functionality and associated benefits within small scale projects, such as community gardens (Firth et al., 2011; Johnson, 2012; Zoellner et al., 2012).

One key function that is often highlighted in the context of small scale projects is the capacity they hold to satisfy the planning system's statutory requirements for community participation and engagement by providing opportunities for co-production (Bovaird, 2007). Another socio-economic function of small scale projects highlighted within the academic literature is the impact of community gardens and allotments when conceptualised as a network or 'system' (Feenstra, 1997, p. 28) or an 'agrifood landscape' (Allen, FitzSimmons, Goodman, & Warner, 2003, p. 61). Thus, the small scale has the potential to complement the social and economic functions attributed to strategic scale projects, and as such this paper seeks to define the role of the *small* scale or *community-scale* in green infrastructure planning as a complementary scale of green infrastructure delivery, management and maintenance.

Exploring the community-scale in green infrastructure planning

Aim

Building on a previous study by Jerome (2012) that indicates there are different categories of groups who are engaged in the voluntary activity of the creation, management and maintenance of small scale green spaces at the local level, the aim of this study is to define in more detail the characteristics of community-scale green infrastructure activity. To this end, the following two main research questions will be addressed in this paper:

- What role do voluntary groups play in creating, managing and maintaining green space at the local level?
- What types of voluntary group are involved in creating, managing and maintaining green space at the local level?

Theoretical background

It is significant to bring focus to the capacity of the *community*-scale to deliver *strategic* value and respond to need as for many years investment in green infrastructure has been defined by the contribution of public sector agencies including: local authorities, The Environment Agency, British Waterways, The Forestry Commission and Natural England (Leeds City Region Green Infrastructure Strategy, 2010). Furthermore, by contributing a richer understanding of activity at the community-scale, we can extend the current academic literature which has tended towards a narrower picture of activity at the local level: one further minimised because of an assumption towards homogeneity of practice, with a predominant focus on community gardens and food-growing initiatives (Firth et al., 2011; Johnson, 2012; Wakefield, Yeudall, Taron, Reynolds, & Skinner, 2007; Yotti Kingsley & Townsend, 2006; Zoellner et al., 2012).

Methods

Case description and sampling

This paper draws on Ph.D research conducted at the University of Liverpool (2011–2015) to contribute to understanding of green infrastructure activity at the community-scale. The broader thesis is concerned with understanding the key factors and forces at play in the longevity and resilience of community-scale green infrastructure activity. This paper is reporting on one aspect of the research study, which is defining community-scale green infrastructure, both in terms of its activity focus and observable approaches to governance (Figure 1). For this aspect, an initial desk-based search of green infrastructure activity was conducted, recording both current and historical groups and projects. This involved drawing on data available from formal sources including project websites and funder websites such as the Big Lottery 'Local Food' archives, plus various online social networks, including 'Project Dirt' and Facebook. The sample area for data collection was the geographical boundary of The Mersey Forest, an official partner of the research study. The Mersey Forest covers a 500 square mile area in the north-west of England. It spans urban, urban fringe and rural land use (The Mersey Forest, 2014); hence, it was possible

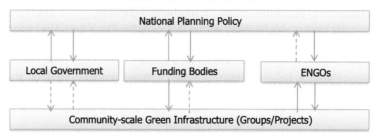

Figure 1. Conceptual framework diagram showing the multi-directional engagement in community-scale green infrastructure activity, and highlighting the gaps in understanding between community-scale green infrastructure (groups and projects) and those responsible for facilitating activity through funding (funding bodies) and access to land (local government).

to explore activity being led by individuals and community groups working in a variety of settings such as street-scale plots, local parks and woodlands. Groups and projects who currently or historically had taken an active role in designing, delivering, managing and maintaining green infrastructure were recorded. It was necessary to limit the sample size however; therefore groups which had discontinued their activities by 2008, the launch date of the Big Lottery 'Local Food' project, were not recorded. The rationale for this decision reflects the theoretical background to the project, which shows that a large number of groups were established as a consequence of this new and significant source of funding for green infrastructure projects at the local level.

Results and discussion

The desk-search returned a population of 300 groups and projects, which were active beyond 2008, or were still currently active in The Mersey Forest area. By organising the groups and projects thematically, it was possible to create a typology of four distinct types: *formal group (active), formal group (inactive), informal group* and *project.* Across the four types it was possible to identify a range of common *benefits,* including conservation and education, health and well-being, food growing, social cohesion, and regeneration. These benefits were observable in varying degrees across the types; and there was often a noticeable correlation between activity focus and benefit/s. This finding builds on Firth et al.'s (2011) typology of small scale green infrastructure which prioritises discussion of variations across groups and projects primarily by *activity.* In contrast, the typology presented here broadens the thematic variation to include categories of *constitutional difference* and *organisational understanding.* The typology therefore describes groups and projects according to activity focus and associated benefits: plus adopted approaches to governance as a key characteristic affecting experiences and outcomes of voluntary groups at the local level.

The theme of governance was primarily utilised to create the four distinct types of community-scale green infrastructure. Firstly, *Groups* distinguish green infrastructure activities which are wholly managed by community volunteers. There are two main types of group: *Formal groups* describe groups who have a formal constitution, such as Friends groups; and *Informal groups* comprise groups who manage and maintain sites of green infrastructure without formal governance structures and systems in place. Secondly, the typology records *Projects* which define green infrastructure activity which is supported by volunteer involvement, but is ultimately managed by employees within an organisation whose aims and objectives are not directly related to green infrastructure. An example of a *project* is a community garden facility developed by a public health organisation to promote well-being benefits associated with access to green space.

By defining voluntary green infrastructure activity at the local level within The Mersey Forest area this study contributes to the research literature by extending a picture of diversity at the community-scale. The typology created shows that many different green infrastructure *functions* and *benefits* are being delivered in response to perceived *need* at the local level. Further, community-scale green infrastructure activity shows a picture of community engagement which is multi-directional: encompassing engagement with a piece of land (over a sustained period of time); engagement with other members of a (place-based/interest-based) community; and engagement between an organisation and its participants. In this context, engagement is understood in terms of *active* engagement, not *passive*, drawing on the work of planning theorists Healey (1992) and Campbell and Marshall (2000), and in particular Arnstein's conceptual framework of active 'participation' (1969). As such, further analysis of community-scale green infrastructure activity, in particular the role of group governance, may also contribute to discussions of citizenship, social responsibility and 'delegated power' (Arnstein, 1969).

Another area of planning theory where community-scale green infrastructure may contribute understanding is the literature which focuses on discussions of place. In the context of this study, 'place' may refer to 'a sense of place' (Carter, Dyer, & Sharma, 2007); 'place attachment' (Green & White, 2007); or a feeling of *connection, belonging,* or *ownership* for/with a physical space outside of home and work (Oldenburg, 2000). As such, *community-scale* can also describe a green infrastructure group or project which has formed around a community with shared or common interests, where a site is created or adopted in response to a green infrastructure need relating to a specific cultural, ethnic, religious or 'other' person-centred need. In this sense, motivational factors (Measham & Barnett, 2008; Seaman, Jones, & Ellaway, 2010) may be *environmental*—to improve the functionality of their own local area; or *social*—in response to physical or mental health needs, social exclusion or isolation, or other social needs.

By observing activity within The Mersey Forest, it was possible to create an understanding of small scale green infrastructure which draws on both these perspectives, highlighting the dual function of social dynamics and place-based factors. One way in which 'connectivity' (Benedict & McMahon, 2006) and 'continuity' (Ahern, 2007, p. 241) across sites of activity were observed was through the creation of 'informal networks' (Littlewood & Thomas, 2010) between individual groups and projects. These networks, which are both spatial and social in character, facilitate opportunities for informal knowledge sharing and resource distribution. As such, *community-scale* green infrastructure describes activity which is both socio-economic and physical (spatial) in its focus and it is precisely the opportunities created for 'larger than site' influence through informal networks of activity which provides the underpinning for the creation of an evolving definition of community-scale green infrastructure activity.

For example, the Friends of Everton Park group in North Liverpool averted a significant impact on their capacity after an arson incident in 2014 by being able to draw on an informal network of green infrastructure-focused community groups in Liverpool to organise a donation of plants and tools. This may be understood conceptually as an example of *bonding* social capital (Firth et al., 2011; Putnam, 2000), describing peer-to-peer partnership across place-based or common-interest networks, or 'collective action' (Bovaird, 2007) which in this case proved to be a key influence on their green infrastructure objectives and retention of volunteers. In another example, The Friends of Furey Wood in Cheshire had a track record of 10 years providing volunteering opportunities. However, after the departure of two key volunteers, their activities discontinued. There were two other volunteer-led woodland groups within 2 km of the Furey Wood site, and although the three groups shared common interests including approaches to working and support from the same local authority Countryside Ranger, for the Friends of Furey Wood the role of informal networks on capacity to adapt to change proved less influential than the role of individuals within the group.

The experience of these two groups helps to advance a theoretical understanding of community-scale green infrastructure activity within a local and place-based or relationship-centred context. Moreover, although there is extensive literature highlighting the importance and challenges associated with community engagement in planning for the built environment in both academic (Arnstein, 1969; Campbell & Marshall, 2000; Healey, 1992) and practitioner (CABE Space, 2009) literature, less attention is shown to the role of community management and maintenance of green infrastructure, with research focusing on participation (Mayer et al., 2012), volunteer motivation (Measham & Barnett, 2008; Seaman et al., 2010) and 'use and enjoyment of the outdoors' (Natural England, 2015). Therefore, by exploring the ways in which individuals and communities organise themselves and create opportunities for the co-production (Bovaird, 2007) of green infrastructure function/s and benefit/s at the community-scale, we can strengthen the conceptual argument that small scale green infrastructure sites provide opportunities for *active* engagement in creation or enhancement of the local environment.

Conclusion

Community-scale green infrastructure describes a complementary scale of green infrastructure planning and delivery: an 'informal network' of *micro* activity to supplement the strategic level interventions which characterise activity at the landscape scale. Community-scale green infrastructure activity is delivered through the efforts of voluntary groups and volunteers. Its green infrastructure *function/s* and *benefit/s* relate to particular communities of place or interest, and can therefore be understood as responding equally to geographic *need* and social *need*. The research being conducted in The Mersey Forest area of North West England indicates a plurality of actors and activities delivering green infrastructure at the local level; challenging the picture of homogeneity in the recent academic literature evaluating the social impact of community gardening and food-growing initiatives (Firth et al., 2011).

The overall picture of the community-scale is thus one of vibrancy and creativity, characterised by groups and projects of various size, membership, and activity focus. In addition, groups and projects are shown to differ in their governance structures, ranging from legally constituted groups to ad hoc informal groups. Through the utilisation of 'informal networks' and social capital, groups and projects develop a more responsive strategy to adapt outputs and outcomes, multi-functionality and associated green infrastructure benefits, to changing local needs, both social and environmental. A more detailed analysis of the factors and forces affecting resilience at a group or project level would be beneficial to substantiate this argument and establish whether the *community-scale* at which these voluntary enterprises are engaged in green infrastructure planning may in fact be the source of their strength.

Disclosure statement

No potential conflict of interest was reported by the author.

Funding

This work was supported by Economic and Social Research Council.

References

Ahern, J. (2007). Green infrastructure for cities: The spatial dimension. In V. Novotny & P. Brown (Eds.), *Cities of the future: Towards integrated sustainable water and landscape management*. London: IWA Publishing.

Alker, J. (2015). We need to shout louder about the business case for green infrastructure. In *Business green*. Retrieved October 19, 2015, from http://www.businessgreen.com/bg/opinion/2397381/we-need-to-shout-louder-about-the-business-case-for-green-infrastructure

Allen, P., FitzSimmons, M., Goodman, M., & Warner, K. (2003). Shifting plates in the agrifood landscape: The tectonics of alternative agrifood initiatives in California. *Journal of Rural Studies, 19*, 61–75.

Arnstein, S. R. (1969). A ladder of citizen participation. *Journal of the American Institute of Planners, 35*, 216–224.

Benedict, M. A., & McMahon, E. (2006). *Green infrastructure [electronic book]: Linking landscapes and communities*. Washington, DC: Island Press.

Bovaird, T. (2007). Beyond engagement and participation: User and community coproduction of public services. *Public Administration Review, 67*, 846–860.

CABE Space. (2009). *Open space strategies: Best practice guidance*. Retrieved July 22, 2015, from https://www.designcouncil.org.uk/sites/default/files/asset/document/open-space-strategies.pdf

Campbell, H., & Marshall, R. (2000). Public involvement and planning: Looking beyond the one to the many. *International Planning Studies, 5*, 321–344.

Carter, J., Dyer, P., & Sharma, B. (2007). Dis-placed voices: Sense of place and place-identity on the Sunshine Coast. *Social & Cultural Geography, 8*, 755–773.

EC European Commission. (2012). Green infrastructure studies. *Green infrastructure implementation and efficiency*. Retrieved March 22, 2015, from http://ec.europa.eu/environment/nature/ecosystems/studies.htm#implementation

Feenstra, G. W. (1997). Local food systems and sustainable communities. *American Journal of Alternative Agriculture, 12*, 28–36.

Firth, C., Maye, D., & Pearson, D. (2011). Developing 'community' in community gardens. *Local Environment, 16*, 555–568.

Green, A. E., & White, R. J. (2007). *Attachment to place—Social networks, mobility and the prospects of young people*. York: York Publishing Services.

Healey, P. (1992). A planner's day: Knowledge and action in communicative practice. *Journal of the American Planning Association, 58*, 9–20.

Jerome. (2012). *MRes dissertation: Common ground: Enabling social capital through community gardening. An exploratory case study for Liverpool's green infrastructure professionals*. Liverpool: Department of Geography and Planning, University of Liverpool.

Johnson, S. (2012). Reconceptualising gardening to promote inclusive education for sustainable development. *International Journal of Inclusive Education, 16*, 581–596.

Leeds City Region Partnership. (2010). *Leeds city region green infrastructure strategy*. Oxford: LDA Design. Retrieved July 21, 2015.

Littlewood, S., & Thomas, K. (2010). A European programme for skills to deliver sustainable communities: Recent steps towards developing a discourse. *European Planning Studies, 18*, 467–484.

Mazza, L., et al. 2011. *Green infrastructure implementation and efficiency*. Final report for the European Commission, DG Environment on Contract ENV.B.2/SER/2010/0059. Brussels: Institute for European Environmental Policy.

Measham, T. G., & Barnett, G. B. (2008). Environmental volunteering: Motivations, modes and outcomes. *Australian Geographer, 39*, 537–552. Retrieved July 23, 2015, from http://dx.doi.org/10.1080/00049180802419237

The Mersey Forest. (2012). *Liverpool green infrastructure strategy*. Retrieved July 22, 2015, from http://www.ginw.co.uk/liverpool/Technical_Document.pdf

The Mersey Forest. (2013). *Liverpool City Region and Warrington Green Infrastructure Framework*. Retrieved September 19, 2014, from http://www.merseyforest.org.uk/our-work/green-infrastructure/liverpool-city-region-green-infrastructure-framework/

The Mersey Forest. (2014). *The Mersey forest plan*. Retrieved October 28, 2015, from http://www.merseyforest.org.uk/about-the-mersey-forest/plan/

National Research Council. (2004). *Valuing ecosystem services: Toward better environmental decision-making*. Washington, DC: National Academies Press.

Natural England and Land Use Consultants. (2009). *Experiencing landscapes: Capturing the cultural services and experiential qualities of landscape*. Retrieved October 19, 2015, from http://www.naturalengland.co.uk

Oldenburg, R. (2000). *Celebrating the third place: Inspiring stories about the 'Great Good Places' at the heart of our communities*. New York, NY: Marlowe & Company.

Peel Land and Holdings. (2015). Webiste for *Wirral Waters*. Retrieved October 19, 2015, from http://www.wirralwaters.co.uk/

Putnam, R. (2000). *Bowling alone: The collapse and revival of American community*. New York, NY: Simon Shuster.

Seaman, P. J., Jones, R., & Ellaway, A. (2010). It's not just about the park, it's about integration too: Why people choose to use or not use urban greenspaces. *International Journal of Behavioral Nutrition & Physical Activity, 7*, 78–86.

Wakefield, S., Yeudall, F., Taron, C., Reynolds, J., & Skinner, A. (2007). Growing urban health: Community gardening in South-East Toronto. *Health Promotion International, 22*, 92–101.

Yotti Kingsley, J., & Townsend, M. (2006). 'Dig in' to social capital: Community gardens as mechanisms for growing urban social connectedness. *Urban Policy & Research, 24*, 525–537.

Zoellner, J., Zanko, A., Price, B., Bonner, J., & Hill, J. L. (2012). Exploring community gardens in a health disparate population: Findings from a mixed methods pilot study. *Progress in Community Health Partnerships: Research, Education and Action, 6*, 153–165.

Common economic oversights in green infrastructure valuation

Alexander Whitehouse

ABSTRACT

Valuation toolkits are supposed to make the valuation of green infrastructure benefits a more simple and accessible task, but there are many barriers which stand in the way of them achieving this. This paper discusses some common economic oversights, associated with double counting, distinct types of value and additionality, which valuation toolkits must avoid to ensure their output is dependable. From this discussion, guidance emerges to help users of valuation toolkits interpret valuation estimates in a meaningful way, and it is concluded that with a good awareness of environmental economic principles, the majority of common issues can be avoided. This conclusion, however, raises the question as to just how accessible a process as complex as green infrastructure valuation can be made to be and places emphasis on the role of toolkit developers who must endeavour to remove as much ambiguity as possible from their methodologies.

Introduction

For many, the time and expertise required to conduct bespoke GI (green infrastructure) valuation places it beyond their means. The propagation, in recent years, of 'valuation toolkits' is a response to this problem, making valuation estimates for a wide range of GI benefits accessible to those who claim no capacity for environmental economics and hold only basic information about the GI feature that composes the valuation subject. Accuracy and transparency are vital attributes of such toolkits, aimed as they are at those deficient in time and/or expertise, as the likely traits of their users combine to make the identification of flaws unlikely (Bagstad et al., 2013). It is, therefore, consequential that a recent review commissioned by Natural England of nine valuation toolkits judged five of the toolkits in question to be not fit for use (see Ozdemiroglu et al., 2013).

For the benefit of valuation toolkit users, it is important to discuss some of the prevalent fallacies which cause toolkits to generate inaccurate or misleading output. This paper aims to do just that, covering the themes of double counting, the distinction between economic impact and economic value, and additionality—common oversights in all forms of economic analysis, which are particularly important within environmental valuation. The paper discusses the nature of each issue in turn and extends to the consideration of what both developers and users of toolkits can do to mitigate the effects of such issues.

Double counting

Double counting occurs when the value of a good or service is erroneously counted more than once within the same valuation exercise (Sloman et al., 2013). That the risk of double counting within the valuation of GI benefits is augmented by the way in which these benefits have traditionally been classified is well documented (see Boyd & Banzhaf, 2007; Fisher & Kerry Turner, 2008; Wallace, 2007), and there is little that can be added to this discussion. There has been significantly less discussion, however, as to the way in which the risk of double counting within GI valuation is affected by the mixture of valuation techniques employed. Revealed preference techniques—chiefly hedonic pricing and travel cost—as well as stated preference techniques such as contingent valuation carry a particularly high risk of facilitating double counting when used as part of a multi-technique approach, which is a concern as they are among the most commonly employed techniques within valuation toolkits. Fu et al. (2011) do highlight the risk of using a blend of these three techniques themselves, but the fact is that the use of any of these three techniques as part of a multi-technique valuation methodology can be problematic, and this issue requires further elaboration.

Whilst revealed and stated preference are effective ways of linking broad values with GI features, using these techniques to isolate the value of individual benefits provided by GI features is difficult (LNE, 2013).

This problem is most acute with hedonic pricing where property market data are used to infer values for GI. Through hedonic pricing, it has been established that, particularly in urban areas, property prices tend to rise as proximity to GI increases, and the reason for this is rooted in a range of GI benefits including green views, recreational opportunities, clean air, noise reduction, shelter, and mental and physical health gains (Garrod & Willis, 1999). Thus, if a toolkit were to value an individual GI benefit such as green views using hedonic pricing, it would inevitably, and perhaps unwittingly, capture elements of the value of numerous other GI benefits. If any of these other benefits were valued separately within the same toolkit, or as part of the same valuation exercise, then double counting would be likely. Though research has been done into the use of spatial data and regression to discount the effects of a wide range of influences on property value, providing a clearer picture of the effect of individual GI benefits (see Geoghegan et al., 1997; Won Kim et al., 2003), the sheer number of factors that need to be taken into account to perfect this approach makes it difficult to claim perfect isolation of any one GI benefit (Kong et al., 2007).

As the use of revealed and stated preference techniques is so common within GI valuation (see LNE, 2013; Merlo & Croitoru, 2005), it would be obstinate to advise against the use of toolkits which include such techniques altogether. It should be recognised though, that when toolkits do employ these techniques to infer values for individual GI benefits, such a condition raises the likelihood of double counting. By identifying such conditions which are conducive to double counting, it becomes possible for users to interpret the output of valuation toolkits in a more meaningful way.

Economic impact and economic value

Being the two most commonly monetised measures found within GI valuation, it is essential that the distinction between economic impact and economic value be recognised (Sunderland, 2012). Although the issues which can arise when this important distinction is disregarded are quite simple in terms of their causes and countermeasures, discussion of this subject within GI valuation literature is scarce, as is its recognition within valuation toolkits.

Economic impact is an observable effect of an event, policy or decision, measured in terms of a change in revenue, value-added, profit, wages, jobs or property prices. Economic value, on the other hand, is a monetary expression of the total extent to which an event, policy or decision is appreciated by an individual, often described as an individual's willingness to pay (Weisbrod & Weisbrod, 1997). There is certainly an overlap between impact and value, as the extent to which an individual appreciates something (value) will likely be influenced by their understanding of the effect of that something on the economy (impact), but this overlap is imperfect for two important reasons. Firstly, an individual's understanding of something's effect on the economy is likely to be flawed, meaning that the value they place on this effect will not be accurate. Secondly, an economic value is influenced by numerous cultural, spiritual and aesthetic factors—such as enhanced enjoyment, satisfaction, community spirit or religious conviction—which have no observable effect on revenue, value-added, profit, wages, jobs or property prices (Sunderland, 2012; Weisbrod & Weisbrod, 1997).

Having a clear understanding of this distinction is of vital importance as the benefits targeted by a single valuation toolkit often comprise a mixture of both economic impacts and economic values (Ozdemiroglu et al., 2013). If in such a case all of the monetary values generated were labelled as economic impacts and added together to form an overall value, the results would be misleading as they would imply that all of the overall value quoted would be realised as observable economic effects, whilst in reality, a portion of this overall value would represent intangible cultural, spiritual and aesthetic factors. A failure to recognise this distinction has been observed in numerous existing toolkits, including those covered by Natural England's valuation tool assessment (see Ozdemiroglu et al., 2013), and this brings about a scenario in which it is likely that users will unknowingly generate unsound results. The solution is simply to return to the definitions of economic impact and economic value, and to categorise the GI benefits valued within a toolkit accordingly, ensuring that the values in each category are kept separate. The lack of such measures within valuation toolkits is an indication that a heightened awareness of basic economic principles is called for within the field of GI valuation.

Additionality

The final economic fallacy discussed here relates to the failure of many GI valuation exercises to demonstrate additionality. This problem materialises when valuation is focussed, as is common, on 'GI interventions', where new GI is planned and the valuation aims to demonstrate its impact. In such cases, what is most important is not the stand-alone value of the GI in question, but the value that it would add to the baseline, also known as additionality (García-Amado et al., 2011; Heinrich, 2014). Essentially, it is about describing the demand side of the GI benefits equation; that is the amount of a benefit that society will actually use, rather than simply the total that could be supplied (Villa et al., 2009).

The additionality of a GI intervention is dependent upon the characteristics of the baseline. If a baseline scenario is one of GI benefit scarcity, with a high beneficiary population, such as may be found in a sparsely planted city centre, the additionality of new GI would be high, as almost its entire beneficial capacity would be appreciated. On the other hand, if the baseline scenario is one of relatively good benefit provision with a smaller beneficiary population, in and around a leafy suburb for example, the additionality of new GI would be significantly eroded by the presence of alternatives and a limited number of beneficiaries. In the latter case, with a baseline scenario of good benefit provision, new GI may still be appreciated by beneficiaries, but only at the expense of longer established GI, resulting in 'displacement' of the utility of GI, rather than actual additional utility.

The demonstration of additionality is a standard consideration within economic impact assessments, but within GI valuation specifically it has not received a huge amount of attention. From what research has been done it is clear that the starting point for demonstrating the additionality of GI benefits is to identify the beneficiaries themselves, and these vary from one benefit to another (Hein et al., 2006; Naidoo et al., 2007). For benefits like carbon sequestration, the task is simple as, the world's atmosphere being well mixed, the benefit of a tonne of carbon sequestered has no spatial bias, with the benefit of a tonne sequestered by a tree in the UK being felt equally worldwide (Costanza, 2008). In contrast, for a benefit such as airborne pollution removal by vegetation, the beneficiaries would be very local, being only those people in close proximity to the source of the benefit source, whose lives would be improved by the resultant cleaner air (Won Kim et al., 2003). Hence, identification of beneficiaries requires an estimation of the geographical range across which each benefit would be felt (Naidoo et al., 2008; Syrbe & Walz, 2012).

The next step is to look at the physical characteristics of the wider area to identify alternative sources of the same benefits, and establish whether the demand for the benefit is already satisfied (Turner et al., 2010). In the case of carbon sequestration, though there are innumerable sources of sequestration, there is a known excess of atmospheric carbon, so we can be confident that the additional sequestration provided by an additional tree in any location would be satisfying a demand. For a benefit like airborne pollutant removal, however, it is much more complicated, the utility of the pollutant removal capacity of an individual tree being dependent on the air quality in its immediate vicinity, which varies depending upon location, weather conditions and time of day, as well as the capacity of alternative pollutant removal measures, be they other GI features, or grey infrastructure solutions (Farber et al., 2002).

Some promising approaches for demonstrating or justifying additionality are beginning to emerge, with spatial data proving to be of vital importance. The Mersey Forest's GI Mapping Method is a good example of the utility of this 'spatial data' in this respect, drawing upon a wide range of socio-economic spatial data to map the need for certain benefits across an area, and matching this with spatial data on the benefit provision capacity of various types of GI to reveal where need and provision do and do not coincide (see Butlin, n.d.). Other examples of comparable approaches include EcoServGIS (Bellamy & Winn, 2013) and ARIES (Villa et al., 2009). However, going beyond the establishment of where demand for GI benefits is and is not being met, to actually expressing the level of demand through analogue measures, and using them to justify additionality and rule out displacement within the valuation of GI interventions, remains an unexplored frontier.

The message to take away from this is to do with the way in which valuation results are to be interpreted. As adjustments reflecting additionality are absent from all toolkits so far developed, it is down to the user to remain aware that the valuation estimates generated will be gross figures. The mapping approaches referenced above (see Bellamy & Winn, 2013; Butlin, n.d.; Villa et al., 2009) are highly useful tools in their own right, but with no reliable method for adjusting valuation estimates to demonstrate additionality, qualitative analysis represents the best alternative.

Conclusion

Through discussion of the technicalities and consequences of double counting, the ignorance of the distinction between economic impact and economic value and the failure to demonstrate additionality— some of the most common economic fallacies present within environmental valuation—this paper aspires to contribute constructively to the work of those developing and promoting environmental valuation approaches. From this discussion, a number of key messages have emerged.

Many of the issues discussed here are exacerbated by a lack of recognition within toolkits. With regard to double counting, and the distinction between impact and value, simply adjusting the way in which toolkits present their output to ensure overlapping and incompatible values are kept separate, and signposting why this has been done, would significantly reduce the risk of both the generation of inaccurate values and user misinterpretation. Crucially, this is something which could be incorporated into the development of a toolkit, and would require no additional knowledge or expertise on behalf of the user.

The demonstration of additionality within toolkits, or rather lack of, is much more difficult to address. Flagging up the issue to users would aid their capacity to interpret any output, but improving the actual accuracy of the output with respect to additionality is not likely to be possible in the absence of user expertise in areas of ecology, economics and GIS.

The requirement for enhanced user expertise to realise truly robust environmental valuations raises the question as to whether the toolkits available at present can be effective in making a process as inherently complicated as GI valuation accessible to a wider audience. At the very least, emphasis must be placed on the ease with which the output of toolkits can be interpreted by users, and toolkit developers must endeavour to improve this aspect of their products, whilst also striving to minimise actual inaccuracy of output. In this vein, a new direction for research in the wider field is implicated, with the development of quantitative techniques for demonstrating the additionality of GI benefits surely being a worthy undertaking.

Disclosure statement

No potential conflict of interest was reported by the author.

References

Bagstad, K. J., Semmens, D., Waage, S., & Winthrop, R. (2013). A comparative assessment of decision-support tools for ecosystem services quantification and valuation. *Ecosystem Services, 5*, 27–39. doi:10.1016/j.ecoser.2013.07.004

Bellamy, C., & Winn, J. (2013). *EcoServ – GIS version 1 (England only): A wildlife Trust toolkit for mapping multiple ecosystem services. Durham Wildlife Trust.* Retrieved July 27, 2014, from http://www.durhamwt.co.uk/wp-content/uploads/2012/06/EcoServ-GIS-Executive-Summary-Only-WildNET-Jan-2013-9-pages.pdf

Boyd, J., & Banzhaf, S. (2007). What are ecosystem services? The need for standardized environmental accounting units. *Ecological Economics, 63*, 616–626. doi:10.1016/j.ecolecon.2007.01.002

Butlin, T. (n.d.). *A green infrastructure mapping method.* Retreived July 27, 2014, from http://www.greeninfrastructurenw.co.uk/resources/A_Green_Infrastructure_Mapping_Method.pdf

Costanza, R. (2008). Ecosystem services: Multiple classification systems are needed. *Biological Conservation, 5*, 350–352. doi:10.1016/j.biocon.2007.12.020

Farber, S. C., Costanza, R., & Wilson, M. A. (2002). Economic and ecological concepts for valuing ecosystem services. *Ecological Economics, 41*, 375–392. doi:10.1016/S0921-8009(02)00088-5

Fisher, B., & Kerry Turner, R. (2008). Ecosystem services: Classification for valuation. *Biological Conservation, 141*, 1167–1169. doi:10.1016/j.biocon.2008.02.019

Fu, B., Su, C., Wei, Y., Willet, I. R., Lu, Y., & Liu, G. (2011). Double counting in ecosystem services valuation: Causes and countermeasures. *Ecological Research, 26*, 1–14. doi:10.1007/s11284-010-0766-3

García-Amado, L. R., Perez, M. R., Escutia, F. R., Garcia, S. B., & Mejia, E. C. (2011). Efficiency of payments for environmental services: Equity and additionality in a case study from a biosphere reserve in Chiapas, Mexico. *Ecological Economics, 70*, 2361–2368. doi:10.1016/j.ecolecon.2011.07.016

Garrod, G., & Willis, K. G. (1999). *Economic valuation of the environment: Methods and case studies.* Cheltenham: Edward Elgar Publishing.

Geoghegan, J., et al. (1997). Spatial landscape indices in a hedonic framework: An ecological economics analysis using GIS. *Ecological Economics, 23*, 251–264. doi:10.1016/S0921-8009(97)00583-1

Hein, L., Koppen, K. V., de Groot, R. S., & van Lerland, E. C. (2006). Spatial scales, stakeholders, and the valuation of ecosystem services. *Ecological Economics, 57*, 209–228. doi:10.1016/j.ecolecon.2005.04.005

Heinrich, M. (2014). *Demonstrating additionality in private sector development initiatives.* Retrieved July 27, 2014, from http://www.enterprise-development.org/page/download?id=2400

Kong, F., Yin, H., & Nakagoshi, N. (2007). Using GIS and landscape metrics in the hedonic price modeling of the amenity value of urban green space: A case study in Jinan city, China. *Landscape and Urban Planning, 79*, 240–252. doi:10.1016/j.landurbplan.2006.02.013

LNE. (2013). *Waardering van ecosysteemdiensten: Handleiding.* Retrieved July 31, 2014, from http://natuurwaardeverkenner.be/nwv2/download/Finale_hl_waardering_ESD_layout.pdf

Merlo, M., & Croitoru, L. (2005). *Valuing Mediterranean forests: Towards total economic value.* CABI.

Naidoo, R., Balmford, A., Costanza, R., Fisher, B., Green, R. E., Lehner, B., ... Ricketts, T. H. (2008). Global mapping of ecosystem services and conservation priorities. *Proceedings of the National Academy of Sciences, 105*, 9495–9500. doi:10.1073/pnas.0707823105

Ozdemiroglu, E., Corbelli, D., Grieve, N., Gianferrara, E., & Phang, Z. (2013). *Green Infrastructure – Valuation tools assessment.* Retrieved July 31, 2014, from file:///C:/Users/Alex%20Whitehouse/Downloads/NECR126_edition_1%20(6).pdf

Sloman, J., et al. (2013). *Principles of Economics.* Australia: Pearson Higher Education.

Sunderland, T. (2012). *Microeconomic evidence for the benefits of investment in the environment – Review.* Retrieved July 31, 2014, from http://publications.naturalengland.org.uk/publication/32031?category=49002

Syrbe, R., & Walz, U. (2012). Spatial indicators for the assessment of ecosystem services: Providing, benefiting and connecting areas and landscape metrics. *Ecological Indicators, 21*, 80–88. doi:10.1016/j.ecolind.2012.02.013

Turner, R. K., Morse-Jones, S., & Fisher, B. (2010). Ecosystem valuation. *Annals of the New York Academy of Sciences, 1185*, 79–101. doi:10.1111/j.1749-6632.2009.05280.x

Villa, F., et al. (2009, September). *ARIES (Artificial Intelligence for Ecosystem Services): A new tool for ecosystem services assessment, planning, and valuation.* Proceedings of the 11th annual BioEcon conference on economic instruments to enhance the conservation and sustainable use of biodiversity, Venice.

Wallace, K. J. (2007). Classification of ecosystem services: Problems and solutions. *Biological Conservation, 139*, 235–246. doi:10.1016/j.biocon.2007.07.015

Weisbrod, G., & Weisbrod, B. (1997). Measuring economic impacts of projects and programs. *Economic Development Research Group, 10*, 1–11. Retrieved from http://www.edrgroup.com/pdf/econ-impact-primer.pdf

Won Kim, C., Phipps, T. T., & Anselin, L. (2003). Measuring the benefits of air quality improvement: A spatial hedonic approach. *Journal of Environmental Economics and Management, 45*, 24–39. doi:10.1016/S0095-0696(02)00013-X

CONCLUSION

What next for green infrastructure?

Ian Mell, School of Environment, Education & Development (SEED), University of Manchester

Introduction

The previous chapters in this book illustrated the variability of approach, focus and value attributed to green infrastructure in different locations and by different stakeholders. They discussed aspects of landscape ecology, spatial planning and participation, the role of water management in urban areas and the difficulties in situating green infrastructure thinking in one specific socio-economic discourse (Austin, 2014; Sinnett, Smith, & Burgess, 2015). Alternatively, we see a constant evolution of green infrastructure thinking, policy and practice that focusses on the delivery of appropriate and multifunctional benefits in various local, city and even regional contexts (Mell, 2016b). Such a process of change has been described as a positive, as it allows advocates to utilise the principles of green infrastructure to meet local needs, but also as a hindrance as it limits the certainty of functionality or added value that planners or developers require to support investment. Consequently, where green infrastructure has been successfully developed, we see an engagement with each of these issues, which are rationalised into a more holistic and integrated human-environment-centred approach to landscape planning (Andersson et al., 2014).

Successful investment (and the ongoing management) of green infrastructure therefore relies on a combination of thematic, spatial and temporal factors, which influence *what, where, who* and *how* green infrastructure is developed (Benedict & McMahon, 2006). An understanding of each of these issues is critical to our knowledge, and indeed our experiences of green infrastructure as individuals, communities and cities. Moreover, where advocates within government (at all scales), the environment sector, communities and academia have successfully translated these principles into praxis, we see more extensive investment in green infrastructure (Meerow & Newell, 2017). For example, in Milan, Paris, London and Berlin, we see generations of planners, architects and landscapers using green infrastructure to provide multifunctional spaces (Spanò, DeBellis, Sanesi, & Lafortezza, 2015). Furthermore, in cities facing rapid growth such as Shanghai or New Delhi, we are seeing an embryonic understanding of the added social and economic value that green infrastructure can provide to urban liveability (Mell, 2016b; Wang et al., 2014).

Within each of these regions, and more globally, we are seeing a reinforcement of the principles proposed by Benedict and McMahon (2006), Davies, Macfarlane, McGloin, and Roe (2006) and Williamson (2003), namely: *connectivity, access to nature,* the *promotion of multifunctionality* and the *provision of socio-economic and ecological benefits* within a more holistic approach to landscape management. However, how these principles are employed within development discussions, policy formation and implementation varies between cities, regions and nations. In Europe, landscape planning has long held these principles as key development and management objectives (European Commission, 2013). Likewise, in North America the role of connective and accessible landscapes has a clear lineage within development discussions (Little, 1990; McHarg, 1969). Even in China, landscape form has been central to the ways in which people interact and understand the influence and impact of climatic change, landscape characteristics and social meanings of space (Li, Wang, Paulussen, & Liu, 2005). Thus, green infrastructure, as proposed by Davies et al. (2006), has been considered as a re-articulation of existing ideals found within landscape, ecological and social planning but brought together in a more cohesive narrative – old wine in new bottles.

As shown throughout the previous chapters, successful green infrastructure development provides options for planning stakeholders, environmental advocates and for local communities (CABE Space, 2004, 2005b) and supports Louv's (2005) promoting of 'environmental affordances', where interactions are facilitated in both formal and informal ways. Green infrastructure uses this notion to argue for more inclusive design and appropriate management that ensures that the greatest number of needs of a given community are met through investment (Hale & Sadler, 2012). Therefore, when reflecting on who uses, owns or values green infrastructure we must consider what affordances these spaces offer, and potentially how these can be extended.

The promotion of a multifaceted understanding of green infrastructure that engages with a variety of social, economic and ecological issues is therefore crucial in the delivery of sustainable landscape management (Mell, 2014, 2015). Such a process is unfortunately not forthcoming in all locations, as economic or political motivations have been seen to undermine the promotion of green infrastructure. To ensure uptake green infrastructure planning, as discussed throughout the previous chapters, must continue to draw directly on a multidisciplinary, multi-focussed, multi-delivery approach to landscape management (Austin, 2014; Walmsley, 2006). The integration of expertise from architecture, engineering, policymaking and socio-economic urban management should therefore be considered complementary (Ahern, 2013). Where this is achieved, we see innovative and reflective investment in green infrastructure that meets the personal, communal and city-scale needs of a location (Ward Thompson, Aspinall, & Bell, 2010). However, there are several areas that require further investigation, as shown in the previous chapters illustrating ongoing negotiations which are needed to ensure that the focus, delivery and management of green infrastructure is sustainable.

Partnerships, participation and 'green infrastructure' products

To ensure that green infrastructure is (a) used by planners and developers and (b) delivers sustainable urban development there is a need, as outlined by Lennon, Scott, Collier, and Foley (2017) to engage effectively with a broad range of stakeholders. Throughout the literature, there is an ongoing debate reflecting who needs to be engaged in development, who should be engaged and, finally, who is. A reading of this literature indicates that extensive variability exists between projects, stakeholders and locations regarding who is invited to discuss the development (and subsequent management) of green infrastructure. Lennon et al.'s work indicates that there is a need for an ongoing reflection on who is involved to ensure the most appropriate and knowledgeable people are included (Lennon, 2014b). However, an additional examination of consultation from Wilker, Rusche, and Rymsa-Fitschen (2016) argues that this is not a straightforward process, and that conflicts can arise between stakeholders trying to ensure their objectives are met.

Conceptually, Healey's (2006) work on collaborative planning situates this argument stating that, at a basic level, all actors with an interest in a subject should be engaged with the consultation process. Moreover, she goes on to discuss how efforts should be made to include non-specialists. However, as Lennon et al. (2017) and Wilker and Rusche (2013) suggest, this is not necessarily logistically or politically expedient. Whilst communities, third sector organisations and environmental organisations may all have expertise, local knowledge and/or the time to engage with a given green infrastructure development process, their motives may be partial (Zmelik, Schindler, & Wrbka, 2011). Consequently, within the practitioner literature, we see a broad discussion of the role of green infrastructure advocates acting as gatekeepers, facilitators, arbitrators and custodians of the development process, despite calls for other bodies to take charge. This has led to accusations of bias, and a lack of breadth or inclusivity in who and how people are engaged (Liverpool City Council, 2016).

However, where the consultation process is well organised, where stakeholders with expertise and experience of developing green infrastructure are engaged, and where the outcomes of the collaborative process lead to meaningful and appropriate investments, we can identify successful investment (CABE Space, 2004). This can focus on the development and delivery of a green infrastructure strategy, a project or on a programme of landscape investment projects. To achieve these latter successes though requires effective communication of the key benefits of green infrastructure, as discussed by Lennon et al. (2017), and an understanding of how the spatial and temporal evolution of landscape resources influence functionality and value.

Continued spatial and temporal evolution of green infrastructure

Aligning effective communication with development objectives is crucial to the successful investment of green infrastructure. However, there is also a need to ensure that the most up-to-date and informative data are available to help planners identify temporal changes to an environment. In previous chapters, Sanesi, Colangelo, Lafortezza, Calvo, and Davies (2017) discussed the changing spatial dynamics of Milan's urban forests highlighting the added value that investing in such spaces provided for the city of Milan. Their analysis illustrates the ecological and the socio-economic value of investments in large-scale green infrastructure sites. Similarly, Szulczewska, Giedych, and Maksymiuk (2017) argue that through a more refined analysis of green infrastructure needs the city of Warsaw in Poland has been able to promote a more nuanced understanding of both temporal and spatial change. This is an essential process whereby Local Planning Authority (LPAs) and green infrastructure advocates can consider the influence of demographic and economic change on a landscape, and identify where new green infrastructure resources can be used to meet a range of climatic, economic and social needs. The development of green infrastructure strategies in Cambridgeshire in the UK (Cambridgeshire Horizons, 2011; Mell, 2016b), New York and Philadelphia in the USA (New York City Environmental Protection, 2010; Philadelphia Water Department, 2011), and the integration of urban greening into the strategic development plan for Ahmedabad all highlight the value such an understanding (Ahmedabad Urban Development Authority, 2013).

Ensuring that planners, developers, green infrastructure advocates and the public are aware of the constant temporal and spatial reassessment needed for effective management is, however, difficult. As Jerome (2017) notes in this book how people interact with their local environments and what they consider green infrastructure to be can differ significantly between locations. Moreover, in cities where green infrastructure has a historical legacy, for example London or Paris, we see more vociferous discussions of how best to manage the environment, and more visible dissent when green infrastructure is threatened (cf. Liverpool City Council, 2016; Mell, 2016a). We should also be aware that such discussions should reflect upon how water (and specifically storm water) and biodiversity are managed. As Dagenais, Thomas, and Paquette (2017) argue the more we understand about the local context, the more likely we are to design and implement water management practices that are appropriate to a location. This can be at a local, city or even a regional scale but relies on an alignment of the principles of the water cycle and local context (Hellmund & Smith, 2006). Likewise, our promotion of biodiversity needs to discuss the roles of existing landscape features, local context and supporting ecological networks, if we are to site investment appropriately (Farina, 2006). The discussion of stone pine forests in this book is just one example of how spatial variation and changing landscape values have influenced ecological diversity and spread over time (Gasparella et al., 2017).

We therefore require a more reflective approach to development that (a) understands the history of a location, (b) its value to local communities and ecology and (c) that investment is spatially responsive to these changes, which subsequently needs to be reflected in policy and practice. Furthermore, our use of green infrastructure resources needs to support existing infrastructure but should also to ensure that networks are connected within and across the landscape by addressing water, biodiversity and climate change issues at multiple scales (Austin, 2014; Benedict & McMahon, 2006; Sinnett et al., 2015).

The challenge of delivering existing, new and retrofitted green infrastructure

The need to adapt our landscapes to the challenges of changing demographics, climate and economic needs raises the question of how we manage existing green infrastructure resources and where new or retrofitted investments could take place (Austin, 2014; Mell, 2016b). In locations where land values could be considered as prohibitive for investment in green spaces, for instance in Tokyo or Shanghai, researchers are reviewing how the green technologies of roofs, walls and water systems can be used to integrate green infrastructure into high-density areas (Caspersen, Konijnendijk, & Olafsson, 2006; Jim, Lo, & Byrne, 2015; Watanabe, Amati, Endo, & Yokohari, 2008). Such investments have the added value of provided climatic mitigation to heat, rainfall and pollution stresses, as well as providing visible symbols of urban greening (Gill, Handley, Ennos, & Pauleit, 2007). This form of retrofitting is also reported by Swilling (2011), Norton et al. (2015) and Tzoulas et al. (2007) as providing economic, as well as health and well-being benefits through environmental psychology perspectives of visible nature.

Current green infrastructure research also focusses on the structures and barriers within planning and development of delivering new resources. Ensuring that stakeholders are collectively supportive of green infrastructure investment has been shown to be varied in different locations. Roe and Mell (2013) proposed that decision-makers can be subject to a form of 'institutional malaise' illustrating the variability of approaches and support for investment in green and open spaces. Lennon et al.'s (2017) discussion of engagement with multidisciplinary understandings of green infrastructure is Ireland, is one example of this. Moreover, the development of the PlaNYC – New York City's green infrastructure plan (New York City Environmental Protection, 2010) and more recently the formulation of the Atlanta Beltline delivery plan, both highlight the positives of stakeholder involvement in developing retrofitted green infrastructure (Kirkman, Noonan, & Dunn, 2012). Unfortunately, such support for new investment is not universal. For example, within the UK, ongoing austerity measures are placing excessive constraints on city managers who are struggling to manage existing spaces without considering new investments (Mell, 2016a).

One consequence of such fiscal uncertainty is the rise in innovative and 'green-sky' thinking that stakeholders are using to assess how they can integrate green infrastructure into new or existing developments (Liverpool City Council, 2016). This includes discussions of different funding mechanisms, such as Park Trust models or more corporate sponsorship, the promotion of micro-scale interventions that can manage storm water in urban Chicago coordinated by the Center for Neighborhood Technology (CNT) (Center for Neighborhood Technology, 2015), or the role of retrofitting transport infrastructure with additional greening, for example in Shanghai, which can improve the aesthetic and climatic properties of a city (Mell, 2016b; Wang et al., 2014). Thus, we are witnessing a rethinking of Walmsley's (2006) concern that green infrastructure is only developed when all other infrastructure has been delivered. The conversations we are now hearing focus more frequently on the added value that green infrastructure can deliver to people, communities and the economy (Mell, Henneberry, Hehl-Lange, & Keskin, 2013; South Yorkshire Forest Partnership & Sheffield City Council, 2012).

People, places and the socio-economic value of green infrastructure

One aspect of green infrastructure thinking, that is constantly being refined, is the understanding of who it is for and what benefits it provides. Jerome's (2017) discussion of the community scale brings forth an important aspect of this debate illustrating the need to understand local context and how people interact with the landscapes around them. In a significant proportion of the green infrastructure literature, the concept is debated in a more abstract and spatially broader sense (Wolch, Byrne, & Newell, 2014). What Jerome does is rationalise these macro principles to identify consensus for green infrastructure value at the micro-scale. Other authors, including Schmelzkopf (2002), have also reported on the role of community gardens in promoting engagement with the landscape, as well as local identities. However, there remains a limited evidence-base supporting this nuanced assessment of green infrastructure value at the local scale (Lennon, 2014a).

Whilst Jerome (2017) aimed to define our understanding of the community scale, Whitehouse (2017) attempts to work more broadly by identifying the barriers to effective economic valuation of green infrastructure. He discusses the variability endemic in valuation techniques, processes and outcomes, which has been a significant issue in translating the established techniques from built environment economics onto landscape practices. However, the valuation of green infrastructure has become increasingly important as planners, politicians and developers are looking to maximise their return on investments (Irwin, 2002). Unfortunately, the variability of approaches and how each aspect of a green infrastructure resource can be accounted for remains difficult to rationalise. What Whitehouse and other commentators, such as Greed (2011), Sutherland (2012) and Tyrväinen (2001) have done is to ensure that by using economic valuation techniques that green infrastructure is debated in the same conversations as 'grey' or built infrastructure.

Both the qualification of economic value and the definition of green infrastructure benefits at a local, city and even regional scale are becoming increasingly prominent in the academic and practitioner literature (CABE Space, 2005a). Investigations are important as they provide additional evidence of the value of green infrastructure at both a micro (or community) scale and economically at a macro scale (Brown & Raymond, 2007; Mell, Henneberry, Hehl-Lange, & Keskin, 2016). Furthermore, with a growing societal understanding of the links between health, well-being, social/community inclusion, education, environmental behaviour, physical activity and environmental awareness and green infrastructure, we can start to create a development narrative that utilises the scalar variation and economic returns from these resources to promote investment (Beatley & Newman, 2013; Coutts, 2016; Greed, 2011). All of which needs to be considered as part of a people–place–environment discourse where the values of people are discussed in conjunction with an understanding of specific locales (and their sociocultural meanings), and its physical resource base (Mell, 2016b). Where this is achieved, we see investment that is reflective of local needs and supportive of socio-environmental interactions (Sinnett et al., 2015).

What next for green infrastructure?

The future appears positive for green infrastructure planning. We are seeing more projects, with higher design quality meeting personal and communal needs, being developed around the world. We are seeing an increased understanding of how landscape networks, ecosystem services and water resources can be integrated into development. We are also witnessing greater uptake of green infrastructure within political, economic and development conversations (Benedict & McMahon, 2006). Such progress seemed unlikely when green infrastructure was an embryonic notion being discussed in the USA in the late 1990s (Ahern, 2013). Move forward almost twenty years and we are seeing a maturity in how green infrastructure is discussed in Europe and North America, whilst locations in Asia (east and south) are digesting the research at a rapid rate and integrating 'green' thinking into landscape and urban management (Mell, 2016b). We can therefore propose a series of development trajectories for green infrastructure research over the coming years.

First, as with the development of city-wide green infrastructure networks in Liverpool and Belfast, and the Garden by the Bay in Singapore, there is a move towards larger and more ambitious investments. This includes innovative greening technology, city-scale connective links and investment in larger multifunctional spaces (Austin, 2014). Second, as identified in Milan and Warsaw, investment in urban greening will look to balance the existing resource base with the need to retrofit these spaces with further landscaping. Moreover, we are seeing a more extensive use of green infrastructure to improve the aesthetic and ecological quality of infrastructure such as urban highways and derelict industrial spaces, for example with the Atlanta Beltline, Miami Underline and Duisburg-Nord Landschaftspark. Third, we are starting to witness a political use of green infrastructure to meet a range of socio-economic and ecological issues. In Philadelphia, New York and Chicago, strong political support has seen investment in water management, biodiversity projects and environmental awareness programmes (Mell, 2016b; Philadelphia Water Department, 2011). In the UK, England's Community Forests have acted as key environmental custodians in many low-income areas promoting green infrastructure in local political discussion (England's Community Forests, 2004).

Each of these areas provides scope for advocates to use green infrastructure within, across and between our urban areas and supports investment in multifunctional landscape. The only question remaining is therefore what level of investment is achievable? Many commentators are reluctant to facilitate successful green infrastructure delivery due to the uncertainties of political cycles, communal needs and financial instability; however, the catalogue of evidence linking health, well-being, economic and environmental benefits to investment in green infrastructure could be considered overwhelming. It therefore falls to advocates across different disciplines to continue making the political, economic, social and environmental case for green infrastructure, which will translate into successful delivery.

Bibliography

Ahern, J. (2013). Urban landscape sustainability and resilience: The promise and challenges of integrating ecology with urban planning and design. *Landscape Ecology*, *28*(6), 1203–1212.

Ahmedabad Urban Development Authority. (2013). *Draft Comprehensive Development Plan 2021 (Second Revised)*. Ahmedabad: Ahmedabad Urban Development Authority.

Andersson, E., Barthel, S., Borgström, S., Colding, J., Elmqvist, T., Folke, C., & Gren, A. (2014). Reconnecting cities to the biosphere: Stewardship of green infrastructure and urban ecosystem services. *Ambio*, *43*(4), 445–53.

Austin, G. (2014). *Green infrastructure for landscape planning: Integrating human and natural systems*. New York: Routledge.

Beatley, T., & Newman, P. (2013). Biophilic cities are sustainable, resilient cities. *Sustainability*, *5*(8), 3328–3345.

Benedict, M. A., & McMahon, E. T. (2006). *Green infrastructure: Linking landscapes and communities. Urban Land* (Vol. June). Washington DC: Island Press.

Brown, G., & Raymond, C. (2007). The relationship between place attachment and landscape values: Toward mapping place attachment. *Applied Geography*, *27*(2), 89–111.

CABE Space. (2004). *Green space strategies: A good practice guide*. London: CABE Space.

CABE Space. (2005a). *Does money grow on trees?* London: CABE Space.

CABE Space. (2005b). *Start with the park: Creating sustainable urban green spaces in areas of housing growth and renewal*. London: CABE Space.

Cambridgeshire Horizons. (2011). *Cambridgeshire green infrastructure strategy*. Cambridge: Cambridge Horizons.

Caspersen, O. H., Konijnendijk, C. C., & Olafsson, A. S. (2006). Green space planning and land use: An assessment of urban regional and green structure planning in Greater Copenhagen. *Geografisk Tidsskrift-Danish Journal of Geography*, *106*(2), 7–20.

Center for Neighborhood Technology. (2015). Get RainReady with TREES. Retrieved May 27, 2015, from http://rainready.org/sites/default/files/factsheets/RainReady Trees.pdf

Coutts, C. (2016). *Green infrastructure and public health*. Abingdon: Routledge.

Dagenais, D., Thomas, I., & Paquette, S. (2017). Siting green stormwater infrastructure in a neighbourhood to maximise secondary benefits: Lessons learned from a pilot project. *Landscape Research*, *42*(2), 195–210.

Davies, C., Macfarlane, R., McGloin, C., & Roe, M. (2006). *Green infrastructure planning guide*. Anfield Plain: North-East Community Forest.

England's Community Forests. (2004). *Quality of place, quality of life*. Newcastle: England's Community Forests.

European Commission. (2013). *Communication from the commission to the European parliament, the council, the European economic and social committee and the committee of the regions: Green infrastructure (GI)—Enhancing Europe's natural capital*. Brussels: European Commission.

Farina, A. (2006). *Principles and methods in landscape ecology: Towards a science of the landscape*. London: Springer.

Gasparella, L., Tomao, A., Agrimi, M., Corona, P., Portoghesi, L., & Barbati, A. (2017). Italian stone pine forests under Rome's siege: Learning from the past to protect their future. *Landscape Research*, *42*(2), 211–222.

Gill, S. E., Handley, J. F., Ennos, A. R., & Pauleit, S. (2007). Adapting cities for climate change: The role of the green infrastructure. *Built Environment*, *33*(1), 115–133.

Greed, C. (2011). Planning for sustainable urban areas or everyday life and inclusion. *Proceedings of the ICE—Urban Design and Planning*, *164*(2), 107–119.

Hale, J., & Sadler, J. (2012). Resilient ecological solutions for urban regeneration. *Engineering Sustainability*, *165*(1), 59–67.

Healey, P. (2006). *Collaborative planning: Shaping places in fragmented societies, 2nd Edition*. London: Palgrave Macmillan.

Hellmund, P. C., & Smith, D. (2006). *Designing greenways: Sustainable landscapes for nature and people*. Washington DC: Island Press.

Irwin, E. G. (2002). The effects of open space on residential property values. *Land Economics*, *78*(4), 465–480.

Jerome, G. (2017). Defining community-scale green infrastructure. *Landscape Research*, *42*(2), 223–229.

Jim, C., Lo, A. Y., & Byrne, J. A. (2015). Charting the green and climate-adaptive city. *Landscape and Urban Planning*, *138*, 51–53.

Kirkman, R., Noonan, D. S., & Dunn, S. K. (2012). Urban transformation and individual responsibility: The Atlanta BeltLine. *Planning Theory*, *11*(4), 418–434.

Lennon, M. (2014a). Green infrastructure and planning policy: A critical assessment. *Local Environment*, *20*(8), 957–980.

Lennon, M. (2014b). Presentation and persuasion: The meaning of evidence in Irish green infrastructure policy. *Evidence & Policy: A Journal of Research, Debate and Practice*, *10*(2), 167–186.

Lennon, M., Scott, S., Collier, C., & Foley, K. (2017). The emergence of green infrastructure as promoting the centralisation of a landscape perspective in spatial planning—the case of Ireland. *Landscape Research*, *42*(2), 146–163.

Li, F., Wang, R., Paulussen, J., & Liu, X. (2005). Comprehensive concept planning of urban greening based on ecological principles: A case study in Beijing, China. *Landscape and Urban Planning*, *72*(4), 325–336.

Little, C. (1990). *Greenways for America*. Baltimore: The John Hopkins University Press.

Liverpool City Council. (2016). *Strategic Green and Open Spaces Review Board: Final Report*. Liverpool: Liverpool City Council.

Louv, R. (2005). *Last child in the woods: Saving our children from nature-deficit disorder*. Chapel Hill, NC: Algonquin Books.

McHarg, I. L. (1969). New York. *Design with nature*. John Wiley & Sons.

Meerow, S., & Newell, J. P. (2017). Spatial planning for multifunctional green infrastructure: Growing resilience in Detroit. *Landscape and Urban Planning*, *159*, 62–75.

Mell, I. C. (2014). Aligning fragmented planning structures through a green infrastructure approach to urban development in the UK and USA. *Urban Forestry & Urban Greening*, *13*(4), 612–620.

Mell, I. C. (2015). Green infrastructure planning: Policy and objectives. In D. Sinnett, S. Burgess, & N. Smith (Eds.), *Handbook on green infrastructure: Planning, design and implementation* (pp. 105–123). Cheltenham: Edward Elgar Publishing.

Mell, I. C. (2016a). GI management—time to 'let someone else have a go'? *Town and Country Planning, March/April*, 138–141.

Mell, I. C. (2016b). *Global green infrastructure: Lessons for successful policy-making, investment and management*. Abingdon: Routledge.

Mell, I. C., Henneberry, J., Hehl-Lange, S., & Keskin, B. (2013). Promoting urban greening: Valuing the development of green infrastructure investments in the urban core of Manchester, UK. *Urban Forestry & Urban Greening*, *12*(3), 296–306.

Mell, I. C., Henneberry, J., Hehl-Lange, S., & Keskin, B. (2016). To green or not to green: Establishing the economic value of green infrastructure investments in The Wicker, Sheffield. *Urban Forestry & Urban Greening*, *18*, 257–267.

New York City Environmental Protection. (2010). *NYC green infrastructure plan: A sustainable strategy for clean waterways*. New York: New York City Environmental Protection.

Norton, B. A., Coutts, A. M., Livesley, S. J., Harris, R. J., Hunter, A. M., & Williams, N. S. G. (2015). Planning for cooler cities: A framework to prioritise green infrastructure to mitigate high temperatures in urban landscapes. *Landscape and Urban Planning*, *134*, 127–138.

Philadelphia Water Department. (2011). *Green City, Clean Waters: The City of Philadelphia's Program for Combined Sewer Overflow Control*. Philadelphia: Philadelphia Water Department.

Roe, M., & Mell, I. C. (2013). Negotiating value and priorities : Evaluating the demands of green infrastructure development. *Journal of Environmental Planning and Management*, *56*(5), 37–41.

Sanesi, G., Colangelo, G., Lafortezza, R., Calvo, E., & Davies, C. (2017). Urban green infrastructure and urban forests: A case study of the Metropolitan Area of Milan. *Landscape Research*, *42*(2), 164–175.

Schmelzkopf, K. (2002). Incommersurability, land use, and the right to space: Community gardens in New York City. *Urtban Geography*, *23*(4), 323–343.

Sinnett, D., Smith, N., & Burgess, S. (2015). In. D. Sinnett, N. Smith, & S. Burgess (Eds.), *Handbook on green infrastructure: Planning, design and implementation*. Cheltenham: Edward Elgar Publishing.

South Yorkshire Forest Partnership & Sheffield City Council. (2012). *The VALUE Project: The final report*. Sheffield: South Yorkshire Forest Partnership & Sheffield City Council.

Spanò, M., DeBellis, Y., Sanesi, G., & Lafortezza, R. (2015). *Green Surge: Milan, Italy. Case Study City Portrait; part of a GREEN SURGE study on urban green infrastructure planning and governance in 20 European cities.* Bari, Italy: Universita deli Studi di Bari 'Aldo Moro' (UNIBA).

Sutherland, T. (2012). *Microeconomic Evidence for the Benefits of Investment in the Environment—review. Natural England Research Reports, Number 033.* Peterborough: Natural England.

Swilling, M. (2011). Reconceptualising urbanism, ecology and networked infrastructures. *Social Dynamics, 37*, 78–95.

Szulczewska, B., Giedych, R., & Maksymiuk, G. (2017). Can we face the challenge: How to implement a theoretical concept of green infrastructure into planning practice? Warsaw case study. *Landscape Research, 42*(2), 176–194.

Tyrväinen, L. (2001). Economic valuation of urban forest benefits in Finland. *Journal of Environmental Management, 62*(1), 75–92.

Tzoulas, K., Korpela, K., Venn, S., Yli-Pelkonen, V., Kaźmierczak, A., Niemela, J., & James, P. (2007). Promoting ecosystem and human health in urban areas using Green Infrastructure: A literature review. *Landscape and Urban Planning, 81*(3), 167–178.

Walmsley, A. (2006). Greenways: Multiplying and diversifying in the 21st century. *Landscape and Urban Planning, 76*(1–4), 252–290.

Wang, H.-B., Li, H., Ming, H.-B., Hu, Y.-H., Chen, J.-K., & Zhao, B. (2014). Past land use decisions and socioeconomic factors influence urban greenbelt development: A case study of Shanghai, China. *Landscape Ecology, 29*(10), 1759–1770.

Ward Thompson, C., Aspinall, P., & Bell, S. (2010). In C. Ward-Thompson, P. Aspinall, & S. Bell (Eds.), *Innovative approaches to researching landscape and health.* Abingdon: Routledge.

Watanabe, T., Amati, M., Endo, K., & Yokohari, M. (2008). The abandonment of Toyko's Green Belt and the search for a new discourse of preservation in Tokyo's Suburbs. In M. Amati (Ed.), *Urban Green Belts in the twenty-first century* (pp. 21–37). Aldershot, UK: Ashgate.

Whitehouse, A. (2017). Common economic oversights in green infrastructure valuation. *Landscape Research, 42*(2), 230–234.

Wilker, J., & Rusche, K. (2013). Economic valuation as a tool to support decision-making in strategic green infrastructure planning. *Local Environment, 19*(6), 702–713.

Wilker, J., Rusche, K., & Rymsa-Fitschen, C. (2016). Improving participation in green infrastructure planning. *Planning Practice & Research, 31*(3), 229–249.

Williamson, K. S. (2003). *Growing with green infrastructure.* Doylestown, PA: Heritage Conservancy.

Wolch, J. R., Byrne, J., & Newell, J. P. (2014). Urban green space, public health, and environmental justice: The challenge of making cities 'just green enough'. *Landscape and Urban Planning, 125*, 234–244. Retrieved from http://doi.org/10.1016/j.landurbplan.2014.01.017

Zmelik, K., Schindler, S., & Wrbka, T. (2011). The European Green Belt: International collaboration in biodiversity research and nature conservation along the former Iron Curtain. *Innovation: The European Journal of Social Science Research, 24*(3), 273–294.

Index

T - #0194 - 111024 - C124 - 246/174/6 - PB - 9780367892227 - Gloss Lamination